Antarctica - A Key To Global Change

Edited by Masaki Kanao, Genti Toyokuni and Masa-yuki Yamamoto

Published in London, United Kingdom

IntechOpen

Supporting open minds since 2005

Antarctica - A Key To Global Change
http://dx.doi.org/10.5772/intechopen.75265
Edited by Masaki Kanao, Genti Toyokuni and Masa-yuki Yamamoto

Contributors
Masaki Kanao, Takahiko Murayama, Elena Shevnina, Ekaterina Kourzeneva, Mohammad Nuruzzama, Elizabeth Thomas, Dieter Tetzner, Mauro Regi, Marcello De Lauretis, Gianluca Redaelli, Patrizia Francia, Joan Lluis Pijoan, David Badia, Joaquim Porté, Rosa Maria Alsina-Pages, Josep Maso, Agustin Zaballos, Giichiro Ohno, Shinji Otani, Atsushi Ikeda, Stephen Palm, Yuekui Yang, Vinay Kayetha

First published in London, United Kingdom, 2019 by IntechOpen
IntechOpen is the global imprint of INTECHOPEN LIMITED, registered in England and Wales, registration number: 11086078, The Shard, 25th floor, 32 London Bridge Street
London, SE19SG - United Kingdom
Printed in Croatia

British Library Cataloguing-in-Publication Data
A catalogue record for this book is available from the British Library

Additional hard copies can be obtained from orders@intechopen.com

Antarctica - A Key To Global Change
Edited by Masaki Kanao, Genti Toyokuni and Masa-yuki Yamamoto
p. cm.
Print ISBN 978-1-78985-815-0
Online ISBN 978-1-78985-816-7

For EU product safety concerns:
IN TECH d.o.o., Prolaz Marije Krucifikse Kozulić 3, 51000 Rijeka, Croatia, info@intechopen.com or visit our website at intechopen.com.

Meet the editors

Dr. Masaki Kanao obtained his PhD degree from Kyoto University. He is currently working at the National Institute of Polar Research in Tokyo, Japan. He is chiefly interested in Earth's structure and evolution from a seismological point of view. He investigated the polar regions, both in the Arctic and Antarctic, as passive and active seismic sources. Dr. Kanao is also interested in the dynamics and tectonics of Earth's continental lithosphere. The Antarctic continent, as a member of the past Gondwana supercontinent, has been the main focus of his work to reveal the history of the lithospheric evolution. Dr. Kanao has also been studying glacial earthquakes—cryoseismic events—in Greenland and Antarctica associated with environmental changes triggered by global warming. These investigations in the polar regions contributed to the development of global Earth sciences, as well as the Federation of Digital Seismological Network and the International Lithospheric Program, which are all affiliated with the International Association of Seismology and Physics of the Earth's Interior.

Dr. Genti Toyokuni is an assistant professor at the Department of Geophysics, Tohoku University, Sendai, Japan. He received his BSc (2004), MSc (2006), and DSc (2009) degrees from Kyushu University, Japan. He was a part-time teacher of earth science at the National Institute of Technology, Kurume College, Japan, during 2007–2009, and a postdoctoral fellow at the National Institute of Polar Research, Japan, during 2009–2011. His research interests include polar seismology (cryoseismology), seismic waveform analysis, numerical modeling of seismic wave propagation, and seismic tomography. He is a member of the Greenland Ice Sheet Monitoring Network (GLISN) and has been participating in field observations of the Greenland ice sheet every year since 2011 with other US and Japanese GLISN members.

Masa-yuki Yamamoto, PhD, is currently affiliated with Kochi University of Technology (KUT), Japan. He specializes in upper atmospheric physics, developing in-situ and remote-sensing equipment and collaborating with JAXA and NASA, as well as with amateur astronomers and high school students. He majored in geophysics and graduated from the Graduate School of Science, Tohoku University, in 2001. After taking a position as a researcher at the Communications Research Laboratory, Tokyo, he moved to KUT as a lecturer in 2003. He became a professor at the School of Systems Engineering in 2013. His specialty is geophysics, especially atmospheric sensing using infrasound observations and sounding rockets. A minor planet was recently named Masayuki-yamamoto (58184) by T. Seki, who discovered it in 1991.

Contents

Preface

Antarctica is the key to understanding the past and present status of Earth's systems, as well as predicting future facts about our planet as viewed from the polar region. Antarctica, composed of an ice-covered continent at its center and surrounded by the Southern Ocean, has been investigated during the last half century by researchers from a variety of scientific branches, including bioscience, the physical sciences, geoscience, oceanography, and environmental studies, together with the technological domain. This book covers recent developments in the scientific research in and around Antarctica, with a view to monitor current variations in the extreme environment, effects from the remarkable changes in temperature and sea-ice extent, the mass loss of ice sheets and glaciers, and variations in marine and terrestrial ecosystems, including human activities.

Masaki Kanao
National Institute of Polar Research,
Tokyo, Japan

Genti Toyokuni
Department of Geophysics,
Tohoku University,
Sendai, Japan

Masa-yuki Yamamoto
Kochi University of Technology,
Kami-shi, Kochi, Japan

Chapter 1

Introductory Chapter: Antarctica - A Key to Global Change

Masaki Kanao

1. Introduction

"Antarctica" is a keystone to understand the past and present status on the Earth system, as well as to predict future figures of our planet as viewed from the polar region. The Antarctic region, composed by ice-covered continent in its center and surrounding the Southern Ocean, has been gradually investigated by a human being during the last half-century by all kinds of scientific branches: bioscience, physical sciences, geosciences, oceanography, environmental studies, and together with their technological innovative developments. Antarctica occupies a pivotal location to have an understanding of the evolution of the past supercontinents such as Rodinia and Gondwana as well as the present day plate movement among global tectonics. Major location names in the Antarctic region are illustrated in **Figure 1**.

2. Scope of the book

This book covers the topics on recent development of all kinds of scientific researches in and around the Antarctic region, with a viewpoint to monitor the current variations in the extreme environment, affected by remarkable changes in temperature and sea-ice extent, mass loss of ice-sheet and glaciers, variations in marine and terrestrial ecosystem including human activities. Multi-disciplinary and inter-disciplinary approaches are beneficial in bi-polar researches, being focused on the interrelated perspectives which will be needed to understand and quantify their connections with the prediction of future climate system change.

Scientific challenges in the Antarctic include a signature on the presence of the ice sheets, ice shelves, glaciers, and sea-ice in and around the continent. In this concern, particularly the cryosphere system is likely to be influenced by temporal-spatial variations in the surface environment in the Antarctic, and continuous researches of their variability provide direct evidence of climate change. For instance, cryosphere originated seismic signals have been significantly reported and can be classified into several kinds of generating sources: dynamics of ice sheets, ice-caps, sea-ice, oceanic tidal cracks, icebergs and calving fronts of the ice caps, and basal sliding of ice streams and glaciers [1–4].

Since the most exciting initiative in the Arctic region was the International Polar Year (IPY) in 2007–2008, which had been conducted at the 50th anniversary of the International Geophysical Year (IGY 1957–1958). The IPY was a big international program composed of multi-disciplinary science branches: upper atmosphere, meteorology, glaciology, geosciences, oceanography, and biosciences conducted by a significant number of polar scientists involved [5]. The initiative significantly

IntechOpen

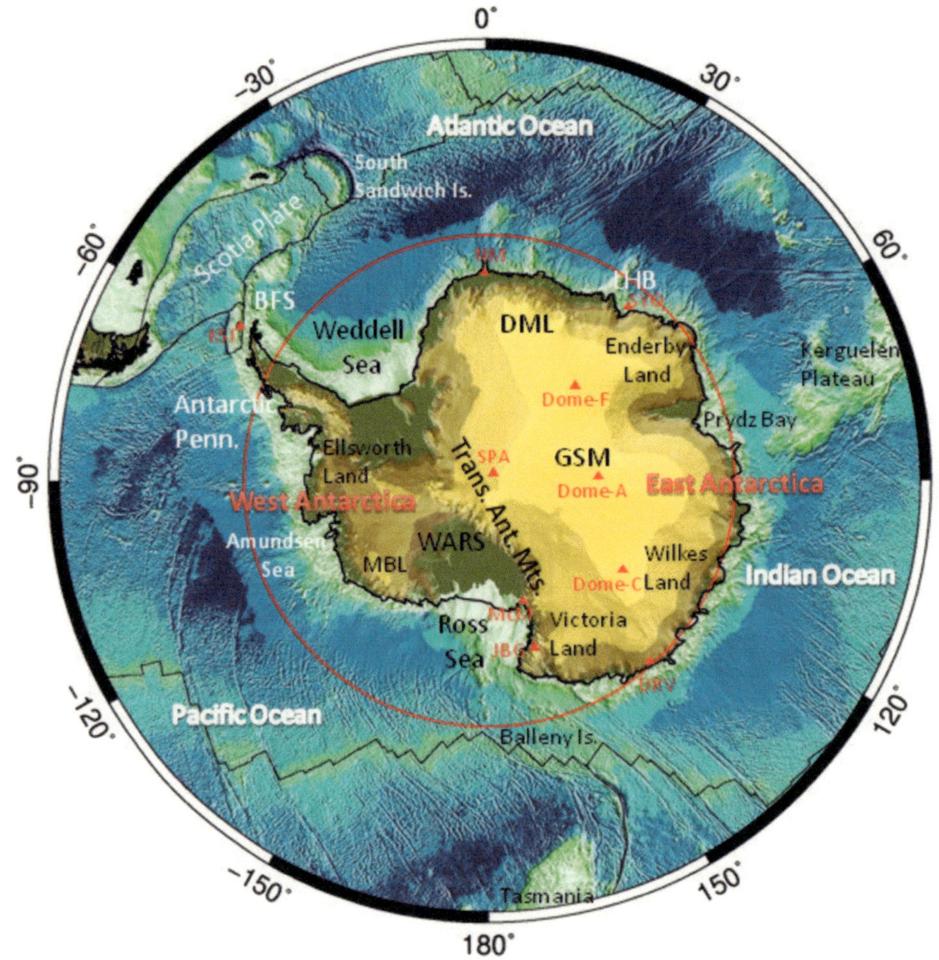

Figure 1.
Surface topography and bathymetry in the Antarctic (ETOPO1, [7]) with major geographic location names treated in this review chapter. Plate boundaries are after [8]. Red solid circle represents the "Antarctic circle" (66.6°S). Abbreviations are as follows. LHB; Lützow-Holm Bay, DML; Dronning Maud Land, GSM; Gambursev Subglacial Mountains; Trans. Ant. Mts.; Trans-Antarctic Mountains, WARS; West Antarctic Rift System, MBL; Marie Byrd Land, BFS; Bransfield Strait. Red solid triangles are the permanent stations. SYO; Syowa Station, NM; Neumayer Station, Dome-F (Fuji), Dome-A (Argus), Dome-C (Charlie), DRV; Dumont D'urville, SPA; South Pole Station, McM; McMurdo Station, JBG; Jang Bogo Station, and KSJ; King Sejong Station.

enhanced the exchange of ideas across nations and scientific disciplines to unveil the status and changes of planet Earth. This kind of inter-disciplinary approach helps us to understand and address grand challenges such as rapid environmental change and its impact on human society.

The recent seismological research achievements in the polar region, for instance, are compiled in the special issue on "Polar Science" [6]. By taking into account the above concerns, however, this book aims to collect many discipline achievements by a significant number of involved projects at the IPY and post era, primarily focusing on surface environmental variations associated with global change such as temperature warming, subglacial lakes distribution, ecosystem dynamics, sea level rise, and melting of the cryosphere.

It is also mentioned that, moreover, the contents in this book intend to appeal not only to the polar scientists but also to all general public who are interested in the present status of the Antarctic region. It is hopeful that this book could provide remarkable knowledge and new understanding regarding current environmental variations and the past Earth's history within the global dynamics. This book could surely attain fruitful information on an advance of frontier researches in the Antarctic region which are currently suffering significant extents by climate change within the global system.

Author details

Masaki Kanao
National Institute of Polar Research (NIPR), Organization of Information and Systems (ROIS), Tachikawa-shi, Tokyo, Japan

*Address all correspondence to: kanao@nipr.ac.jp

References

[1] MacAyeal D, Okal EA, Aster RC, Bassis JN. Seismic and hydroacoustic tremor generated by colliding icebergs. Journal of Geophysical Research. 2008;**113**:F03011. DOI: 10.1029/2008JF001005

[2] Zoet LK, Anandakrishnan S, Alley RB, Nyblade AA, Wiens DA. Motion of an Antarctic glacier by repeating tidally modulated earthquakes. Nature Geoscience. 2012;**5**:623-626

[3] Winberry JP, Anandakrishnan S, Wiens DA, Alley RB. Nucleation and seismic tremor associated with the glacial earthquakes of Whillans ice stream, Antarctica. Geophysical Research Letters. 2013;**40**:312-315

[4] Hammer C, Ohrnberger M, Schlindwein V. Pattern of cryospheric seismic events observed at Ekstrom ice shelf, Antarctica. Geophysical Research Letters. 2015;**42**:3936-3943

[5] Krupnik I, Allison I, Bell R, Cutler P, Hik D, Lopez-Martinez J, et al. Understanding Earth's Polar Challenges: International Polar Year 2007-2008–Summary by the IPY Joint Committee. Edmonton, Alberta: Art Design Printing Inc; 2011. pp. 457-476

[6] Kanao M, Zhao D, Wiens DA, Stutzmann E. Recent advance in polar seismology: Global impact of the International Polar Year—Overview. Polar Science. 2015;**9**:1-4. DOI: 10.1016/j.polar.2014.12.003

[7] Amante C, Eakins BW. ETOPO1 1 arc-minute global relief model: Procedures, data sources and analysis. NOAA Technical Memorandum NESDIS NGDC-24. Boulder, CO, USA: National Geophysical Data Center; 2009. DOI: 10.7289/V5C8276M

[8] Bird P. An updated digital model of plate boundaries. Geochemistry, Geophysics, Geosystems. 2003;**4**:1027. DOI: 10.1029/2001GC000252

Chapter 2

ULF Geomagnetic Activity Signatures in the Atmospheric Parameters in Antarctica

Mauro Regi, Marcello De Lauretis, Gianluca Redaelli and Patrizia Francia

Abstract

The study of the solar wind-Earth's magnetosphere and atmosphere interaction is a topic of great interest. The solar wind energy is transferred to the Earth's environment also through ultralow frequency (ULF, 1 mHz–5 Hz) waves of the geomagnetic field, with higher efficiency at high latitudes where magnetic reconnection processes occur, making the polar cap an important laboratory for these investigations. Several studies suggest that the atmosphere responds to the geomagnetic activity driven by the solar wind, although the interaction processes are not yet completely understood. In this context, the results of recent investigations, showing the coupling on timescales of 1–2 days between geomagnetic ULF activity and the middle-low (h < 50 km) atmosphere in the polar cap, are summarized, based on geomagnetic measurements at Terra Nova Bay, in Antarctica ($\lambda \sim 80°S$) and atmospheric parameters from the reanalysis dataset.

Keywords: ULF waves, solar wind, high-latitude atmosphere, polar cap electrodynamics, energetic particle precipitation, atmosphere processes, cloud microphysics

1. Introduction

In the latest years, the Sun-Earth environment is studied to explain observed physical phenomena in the context of space weather/climate, such as climate changes. It is well known that the Sun continuously transfers its energy to the Earth's environment through radiation and solar wind (SW). Although the Sun's radiation represents the main source affecting the Earth's atmosphere, the SW energy plays an important role during high geomagnetic activity time intervals [1]. On this regard, the magnetosphere-ionosphere represents a complex system able to partially convert SW impacting energy through nonlinearly related physical processes. Such effects are more evident at high latitudes where reconnection processes between interplanetary magnetic field (IMF), carried out by the solar wind, and magnetospheric field occur, making the polar cap an important laboratory to study the SW-atmosphere interactions. Solar wind-driven electrodynamic processes and ultralow frequency (ULF, 1 mHz–5 Hz) waves seem to lead to both diffusion and precipitation processes of energetic electrons in the outer radiation belts, leading also to chemical [2] and microphysical [3] processes in the atmosphere, characterized by different timescales.

The longer-term response, characterized by timescales of several weeks, is usually attributed to the odd nitrogen (NOx) production, due to precipitating energetic electrons, in the mesosphere and lower thermosphere. During the polar winter, NOx can live long enough to be transported downward into the stratosphere where it chemically perturbs the ozone distribution, altering the radiative balance in that region of the atmosphere.

This can in turn affect the overall circulation in the stratosphere, and such changes can propagate to the surface level, eventually leading to detectable changes in surface air temperatures, through dynamical coupling processes occurring on timescales of several weeks (e.g., [2, 4–6]).

Conversely, the shorter-time response (<1 day) of the atmosphere to the SW-magnetosphere coupling processes is probably related to changes in the atmospheric electrodynamics. It can be attributed to electric conductivity variations in the lower atmosphere by ionization mechanisms and/or to changes of the polar cap electric potential induced by SW perturbations [7]. The consequent modulation of the current density which flows from the upper boundary (being as low as 60 km, [8]) through the troposphere to the ground in the global electric circuit (GEC) [3, 9–11] could influence cloud formation through the release of latent heat, which in turn can affect atmospheric dynamics [12].

This work represents a review of our investigations on the experimental observation of the possible short (within $\sim$1 day) timescale response of the atmosphere to the SW dynamics, observed during 2003–2010, which correspond to the solar cycle 23 and the beginning of solar cycle 24. We analyzed the geomagnetic field variations monitored at the Mario Zucchelli station, at Terra Nova Bay (TNB, AACGM latitude λ = 80.01°S and longitude φ = 306.94°E) in Antarctica, and atmospheric parameters at tropospheric and stratospheric heights, provided by ERA-Interim and Monitoring Atmospheric Composition and Climate (MACC) reanalysis archives (http://apps.ecmwf.int/datasets).

The ERA-Interim is a global atmospheric reanalysis dataset, continuously updated in real time (see [13] and references therein). Global atmospheric and surface parameters from 1 January 1979 are available from the surface up to 0.1 hPa as atmospheric fields on model levels and pressure levels, with a temporal resolution of 6 h, and as surface fields with a temporal resolution of 3 h. The data assimilation system used to produce ERA-Interim is based on a 2006 release of the ECMWF Integrated Forecast Model (IFS Cy31r2). The MACC dataset is a global reanalysis dataset of atmospheric composition data, produced by assimilating satellite data into a global model and data assimilation system (see [14] and references therein). The system includes a four-dimensional variational analysis (4D-Var) with a temporal resolution of 12 h analysis window. The ERA-Interim and MACC data, at 1 day resolution used for our studies, have been retrieved from the Meteorological Archival and Retrieval System at ECMWF.

Solar wind parameters and interplanetary magnetic field are monitored by using OMNI data, time-shifted to the bow shock nose (i.e., the subsolar position of the supersonic-to-subsonic transition regions) and collected on CDAWeb (http://cdaweb.gsfc.nasa.gov). Geomagnetic activity was monitored by using a triaxial search-coil magnetometer data, recorded at TNB, at a sampling rate of 1 s.

The atmospheric parameters must be whitened to filter the longer period components (essentially 1 year and 6 months), which would obscure the weak effects produced by the ULF geomagnetic activity [15].

This review is structured as follows: in Section 2, we shortly introduce the interactions occurring between ULF waves and the energetic electrons in the outer radiation belt, leading particle precipitations, as well as polar cap potential difference related with SW parameters; in Section 3, we present the experimental evidence of SW effects on the atmospheric parameters in Antarctica at stratospheric and tropospheric heights; finally, in Section 4, we discuss the estimated timescale response of atmospheric parameters and the possible physical processes involved in SW-atmosphere coupling processes. For greater clarity, we described data analysis and methods in each section.

2. The solar wind-magnetosphere coupling processes at high latitudes

The estimation of the Earth's surface temperature and lower atmosphere energy budget significantly changes due to small amount, distribution, or radiative properties of clouds [16]: therefore, they represent one of the largest sources of uncertainty in predictions of climate change [17]. Even small atmospheric electrical modulations can affect aerosol nucleation processes and cloud condensation nuclei production in troposphere and thus modify cloud properties. In this regard, the polar cap electrodynamics and the energetic particle precipitation seem to be important SW-atmosphere coupling mechanisms, responsible for atmospheric changes on timescales from several weeks to days. It is also known that a global electric current flows in the global electric circuit. It is generated mainly by charge separation in clouds at the tropics and maintains the global ionosphere at a potential of about 250 kV. Variations above and below this value occur in the high-latitude regions due to SW-magnetosphere-ionosphere coupling processes. In this section, we briefly discuss the solar wind-magnetosphere coupling processes, which could produce observable effects in the stratospheric and tropospheric dynamics due to energetic particle precipitation from the outer radiation belt, as well as to the polar cap electrodynamics at high latitudes.

2.1 ULF interaction with relativistic electrons in the outer radiation belts

ULF magnetohydrodynamic waves received particular attention in the past decades [18–23], since they provide a convenient probe of the magnetosphere, by means of ground [24–26] and/or satellites magnetic field measurements [22, 27–30] as well as inspect ground conductivity [31–33].

Generated by a variety of instabilities, ULF waves transport energy throughout the magnetosphere and are observed on the ground as continuous pulsations (Pc, **Table 1**). They can play important roles in the energization and loss of radiation belt particles (see [34] for a review). In particular, ULF waves can interact with the relativistic electrons (>300 keV) magnetically trapped in the radiation belts ($L \sim 5$–7 Re, $\lambda \sim 60$–70°, where L is the McIlwain parameter). In that regions, the charged particle is subject to gyro, bounce, and drift periodic motions (see **Table 1**), each one characterized by different timescales [35–36].

In particular, wave-particle interactions are theoretically predicted [37, 38] because drift and bounce motion frequencies of trapped electrons are in the Pc5 (1–7 mHz) and Pc1-2 (100 mHz–5 Hz) frequency range, respectively. Experimental evidence confirms diffusion/acceleration of energetic electrons by Pc5 magnetospheric waves [39–42] and their precipitations after pitch angle scattering, due to gyro-resonant interaction with electromagnetic ion cyclotron (EMIC) waves [38–41]. Such waves are in the Pc1-2 frequency range and are generated at the magnetic equator by unstable distributions of ring current ions during geomagnetic storms [43]. Moreover, recent investigations show that Pc5 waves have their origin also in the leading edge of the corotating interaction regions (CIR) [42], while the

ULF waves			Magnetically trapped particles		
Pulsation type	Frequency range (mHz)	Period range (s)	Characteristic periodicity (s)		Motion type
			Electrons	Protons	
Pc5	2–7	**150–600**	**10^2–10^3**	10^2	**Drift**
Pc4	7–22	45–150	—	—	—
Pc3	22–100	10–45	—	—	—
Pc2	100–200	5–10	—	—	—
Pc1	200–5000	**0.2–5**	**10^{-1}**	10^0	**Bounce**
—	—	—	10^{-3}–10^{-4}	10^{-1}–10^{-2}	Gyro

Periodicity correspondence between relativistic electrons and ULF waves is marked in bold.

Table 1.
ULF wave classification and the characteristic timescales for the three types of trapped particle motion (see also [37]).

origin of the Pc1-2 waves, observed at ~80° latitude, appears to be due to substorm/storm-related instabilities and, in the dayside, to solar wind compressions of the magnetopause [44]. After conversion into Alfvén (shear) waves, ULF waves propagate along the geomagnetic field lines and can be observed on the ground at high latitudes [45, 46]. An example of energetic electron flux enhancements observed at geosynchronous orbit ($L \sim 6.6$) associated to Pc5 power fluctuations measured on the ground and in the magnetosphere is shown in **Figure 1**, [42]; the electron flux seems to be delayed by ~2 days with respect to pulsation power.

2.2 The roles of relativistic electron precipitation and the polar cap electrodynamics in the global electric circuit model at high latitudes

As discussed by Tinsley and Yu [48], the galactic cosmic rays (GCR) flux is responsible for almost all the production of ionization below 15 km of altitude, which determines the conductivity in that region and at high latitudes. However, the MeV electrons and their associated X-rays produce ionization in the stratosphere and higher troposphere, which can affect the local conductivity (see also [7]). Moreover, due to electric potential difference between the ionospheric layer and the ground, a vertical current density is present:

$$J_z = \sigma E_z. \quad (1)$$

It is directed along the stratosphere-troposphere-ground direction z, and it varies horizontally, due to variations in the local vertical column resistance represented by its resistivity σ and by variations in the local ionospheric potential $\varphi(\mathbf{E})$. While σ is affected by the GCR and precipitating MeV electron fluxes from the outer radiation belt into the atmosphere, $\varphi(\mathbf{E})$ is strongly dependent on SW-magnetosphere coupling process. In particular, the potential increases in the polar cap where geomagnetic field lines map to magnetospheric regions characterized by coupling processes with the interplanetary magnetic field (IMF); because J_z, flowing through clouds in the troposphere, responds to conductivity and potential changes occurring up to 120 km altitude, it is very effective in linking stratosphere and ionosphere with clouds. Recent reviews by Lam and Tinsley [3] and Mironova et al. [7] well discuss the atmosphere response to current density changes near the poles caused by the interaction of the solar wind with the geomagnetic field; such

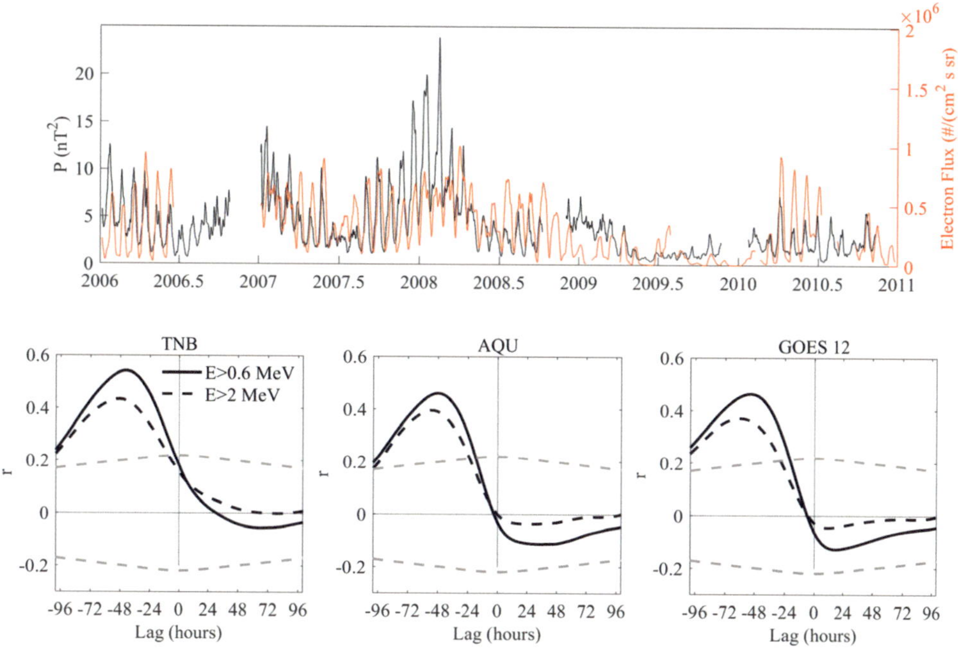

Figure 1.
(Top) The relativistic electron flux (>600 keV) measured by GOES 12 satellite at geosynchronous orbit (red) and the geomagnetic power at TNB (black). (Bottom) The cross-correlation of the >600 keV (solid) and >2 MeV (dashed) electron flux with the Pc5 power at TNB, at low latitude station of L'Aquila AQU, and at GOES 12 during 2007–2008, together with the 95% confidence levels (dashed, gray lines). Figure adapted from [42]. It clearly emerges a higher correlation with the >600 keV electron flux at a shorter-time delay (1.8–2 days) with respect to the >2 MeV electrons (2–2.4 days), with an approximately 9 h difference, probably due to the timescales of the acceleration processes, in agreements with [47].

changes arise in addition to the day-to-day variability in J_z, caused by thunderstorm activity and present at all latitudes.

In particular, the B_y (dawn-dusk) and B_z (south-north) components of the IMF represent the main parameters controlling the electrodynamics in the upper atmosphere (i.e., the ionosphere): B_z is important in the reconnection process in the dayside magnetopause, while B_y affects the latitudinal dawn-dusk electrostatic potential asymmetry [49].

An example of polar cap electric potential in the southern hemisphere is show in **Figure 2**, obtained using the Weimer [50, 51] ionospheric electrodynamic model. In this example, we assigned values of ± 5 nT for B_y and B_z, assuming a solar wind speed of V_{SW} = 450 km/s and a solar wind number density of n_{SW} = 3 cm^{-3}. It is clearly seen that two regions of polar cap potential, characterized by positive (dawnward) and negative (duskward) values of approximately ± 40 kV, are emerging during periods characterized by SW-magnetosphere coupling (i.e., during $B_z < 0$), indicating higher dawn-dusk polar cap potential drop. Conversely, during closed magnetosphere ($B_z > 0$, left panels), lower polar cap potential difference is obtained. It implies that E_z increases at high latitudes during IMF-magnetosphere reconnection ($B_z < 0$) conditions. Furthermore, both σ and E_Z in Eq. (1) can be significantly influenced by geomagnetic activity. In this regard, ULF activity could play an important role in the GEC model, leading to the clouds' physical changes and radiative properties. Indeed, the ULF activity in general is associated to storms and substorms, i.e., to the occurrence of a southward IMF and, at high latitudes, polar cap electric field; therefore, ULF activity can be regarded also as a proxy of the polar cap potential difference [52, 53].

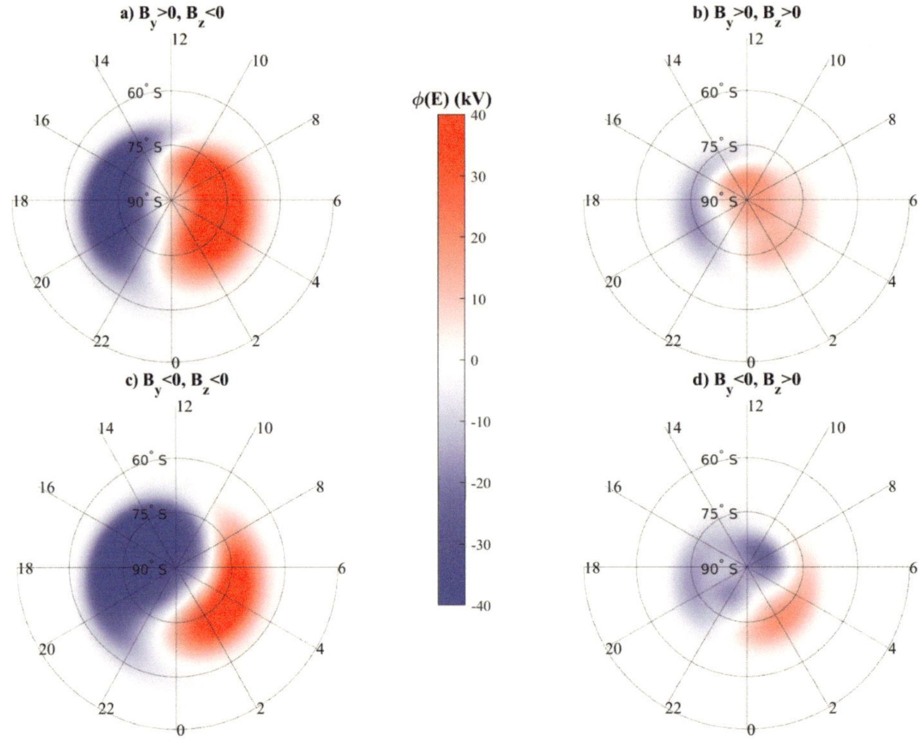

Figure 2.
Computed polar cap electric potential, obtained using Weimer [50, 51] ionospheric electrodynamic model at the equinoctial day 100 of the year 2000 in the southern hemisphere, for open (B_z < 0, left panels) and closed (B_z > 0, right panels) magnetospheric conditions. The meridians are separated by 2 h (MLT) and the geomagnetic parallels by 15°. The morning (afternoon) side corresponds to the right (left) side of each panel.

3. Experimental evidence of signatures of solar wind effects in the atmospheric parameters

In their study, Francia et al. [15] examined surface air temperature measured at the automatic weather station ENEIDE, located at TNB, during 2007–2008, while signatures of ULF activity and polar cap potential difference were found by [54] in the stratosphere and troposphere, using the ERA-Interim reanalysis dataset.

Figure 3 shows the global wavelet (GW), i.e., the time-averaged wavelet [55, 56], separately for summer and winter months, of Pc5 and Pc1-2 powers at TNB and of the polar cap potential difference φ_{cap} [54].

Common power peaks emerge at ~27 days, the Sun synodic rotation period, and its first subharmonics (13.5 and 9 days), in the SW-related parameter φ_{cap} and correspondingly in the ULF activity. In particular, in **Figure 4** signatures of ULF activity and polar cap potential difference in temperature T and zonal wind U at tropospheric and stratospheric heights above TNB are observed. In their work [54], cross-wavelet W_{xy} and wavelet coherence γ^2 [56] are computed between SW-driven and atmospheric parameters; the analysis was supported by the Monte Carlo test for the estimation of significance levels.

In can be seen from **Figure 4** that the correspondence at 27 days is high and statistically significant in both the troposphere and the stratosphere during winter months, while at 13.5 days it is restricted to h < 10 km. During summer, the cross-wavelet W_{xy} for T shows very low values and an ambiguous correspondence at the

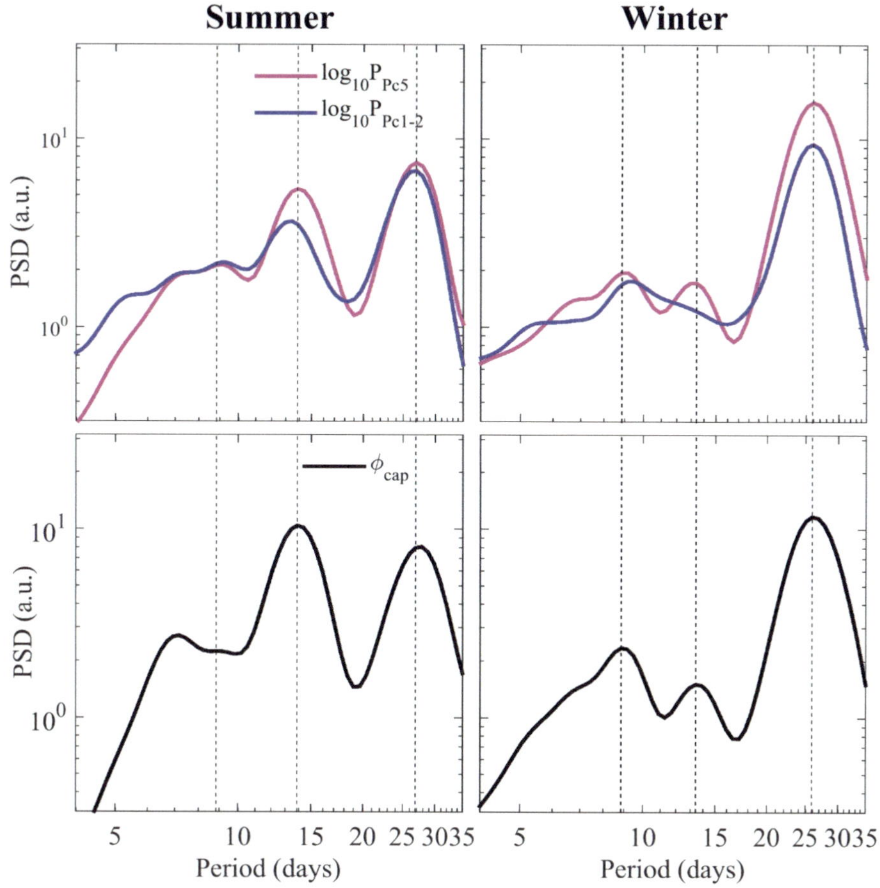

Figure 3.
The normalized global wavelet (GW) analysis during the summer (first column) and winter (second column) in 2007 (top panels); the GW, normalized to the corresponding variances, of the ULF activity indexes logPc5 and logPc1-2 at TNB, and of φ_{cap} (bottom panels). The three first maximum values of the normalized GW are marked with vertical lines. Figure adapted from [54].

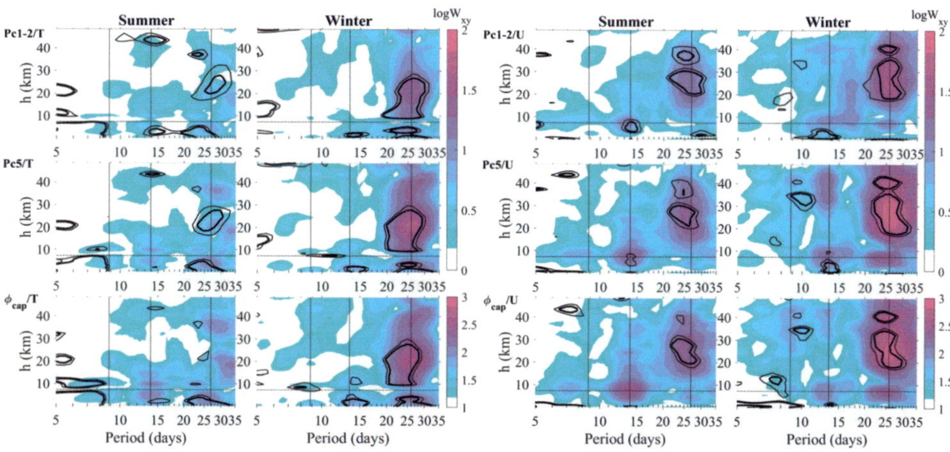

Figure 4.
*Time averaged cross-wavelet amplitude (W_{xy}) between temperature T and zonal wind U with (top) logPc1-2, (middle) logPc5, and (bottom) polar cap potential difference φ_{cap} during summer and winter months in 2007, as a function of period and altitude h. In each panel, the black contour lines mark the 90% (thin) and 95% (bold) significance level of time-averaged wavelet coherence. Vertical lines indicate the periodicities observed in solar wind and geomagnetic activity (**Figure 3**). Figure adapted from [54].*

27 day periodicity for all input parameters. Interestingly, a W_{xy} and γ^2 attenuation can be seen, especially during winter months, near to the tropopause position (~7–10 km). Regarding U, high and significant W_{xy} and γ^2 values are found mostly in the stratosphere, at ~27 and ~13.5 day periodicities, during winter months. During summer the pattern is similar but with lower values.

Analyzing the tropospheric temperature, specific humidity Q and cloud cover (CC) in Antarctica, Regi et al. [57] found further experimental evidence of SW-atmosphere coupling. In particular, the authors analyzed 54 high geomagnetic activity time intervals, by using the superposed epoch analysis with the Monte Carlo test [58].

The results, shown in **Figure 5**, revealed a clear correspondence between Pc1-2 and variations in temperature, specific humidity, and cloud cover. The most significant correspondence is found at latitudes higher than 85°S in both T and Q parameters. Positive variations occurred at the epoch and 1 day after, suggesting that Pc1-2 power affects the specific humidity and the tropospheric temperature within 1 day; negative variations occurred at 2 days after the epoch.

Regarding the cloud cover (zonal mean) composite averages, computed at low (LCC, $h < 3$ km), medium (MCC, $3 < h < 6$ km), and high (HCC, $h > 6$ km) altitudes, (see **Figure 5**, bottom panels), we found that the main effect of the ULF activity on these parameters consists in an increase in the HCC and MCC at the epoch. Moreover, we also found a significant variation at 1 day after and a decrease at 2 days after the epoch.

As discussed in the Introduction, GEC affects atmospheric parameters, such as cloud cover through several proposed microphysical processes (e.g., [3, 10, 48])

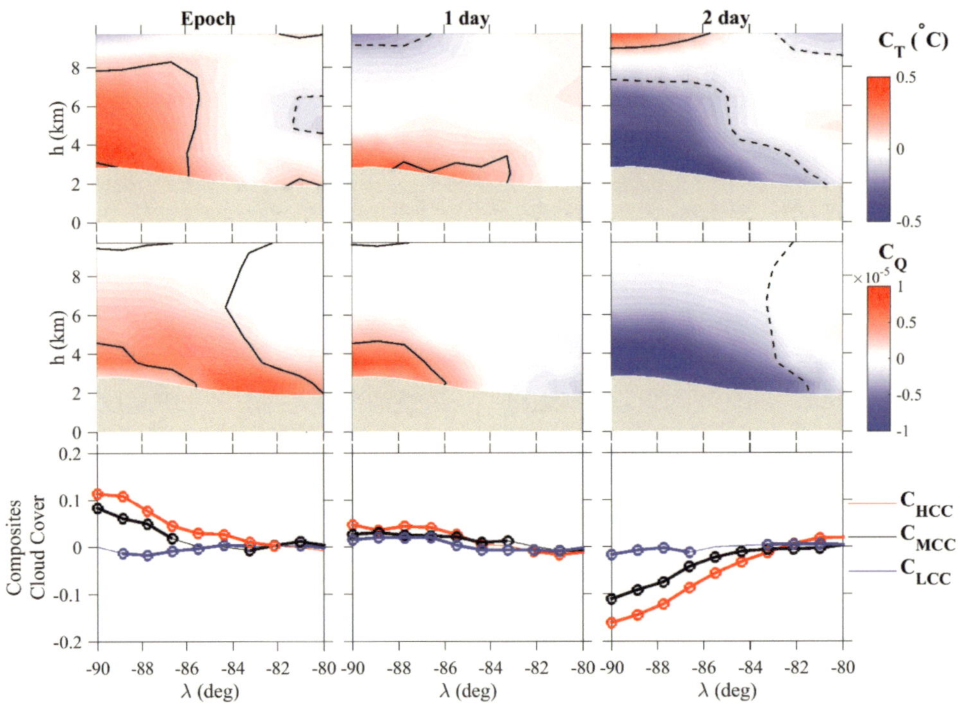

Figure 5.
(Top and middle panels) Composites of the zonal mean of temperature C_T and specific humidity C_Q of 54 selected epochs (corresponding to geomagnetic active periods) as a function of latitude and altitude. Solid and dashed lines indicate the 90% confidence levels. The gray region represents the longitudinally averaged Antarctic icy surface profile. (Bottom panels) Composite mean of cloud cover at different altitudes C_{LCC} (h < 3 km), C_{MCC} (3 < h < 6 km), and C_{HCC} (h > 6 km) as a function of latitude; the bold lines mark composites statistically relevant. Figure adapted from [57].

able to produce changes in the atmospheric parameters, such as temperature. In particular, the accumulation of charge on droplets and aerosol particles, most importantly the interstitial cloud condensation nuclei (CCN) and ice-forming nuclei (IFN), directly affects scavenging rates.

Scavenging is due to collisions between the nuclei and droplets, entailing size-dependent collection of nuclei and changes in their size distribution and overall concentration. It affects a number of microphysical processes, which cause changes in macroscopic cloud properties and in turn partitioning of energy flow in the system. In particular, the charge can increase or decrease the scavenging rates, depending on size, changing the concentrations and size distributions. Size distribution changes in CCN produce size distribution changes in droplets, affecting coagulation, precipitation, latent heat transfer, and cloud cover. Scavenging of ice-forming nuclei by supercooled droplets promotes contact ice nucleation, releasing latent heat. The latent heat changes cause storm invigoration [59] and in winter storms can cause changes in the amplitude of Rossby waves and blocking. Since the typical lifetime of CCN in an air mass can be up to 10 days, the change in their properties can affect also later cycle of evaporation/condensation, flux of latent heat, and the amount of water vapor released into the troposphere.

Figure 6 shows the results of superposed epoch analysis (SEA) conducted by [57] on Q and T at the epoch for different orientations of the IMF. It shows correspondence in T and Q, more evident during the IMF $B_z < 0$ and $B_y > 0$ conditions, i.e., when the interplanetary electric field is efficiently transmitted into the high-latitude ionosphere and the dawn-dusk polar cap potential difference φ_{cap}, due to $E_y = -V_{SW}B_z$, as well as the vertical $E_z = V_{SW}B_y$, increases (see also **Figure 2**). In particular, the vertical electric field during southward IMF conditions can be directly transferred in the ionospheric polar cap in the southern hemisphere,

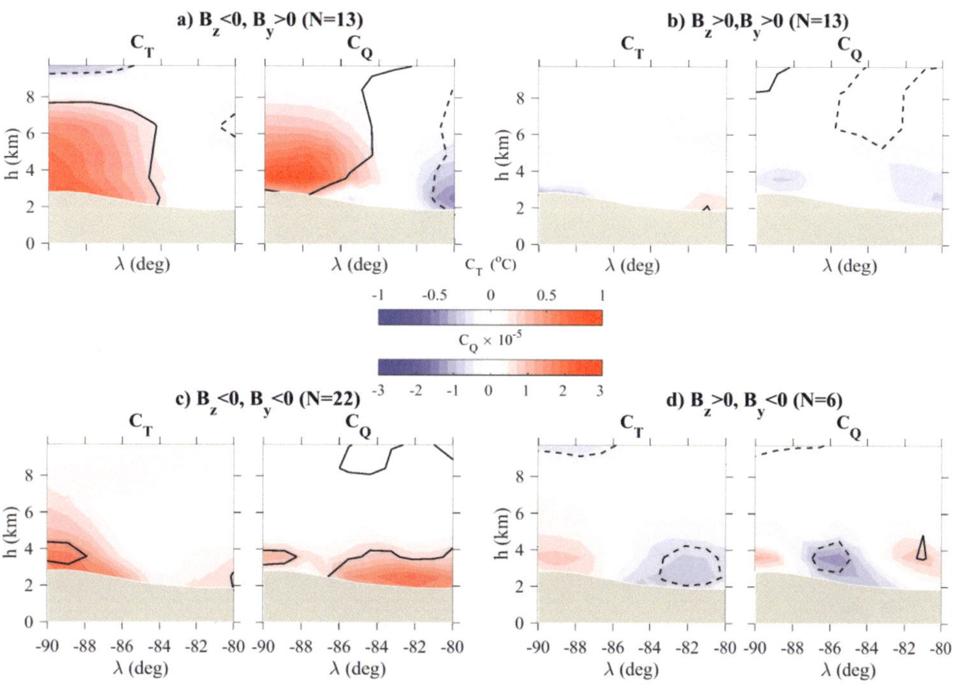

Figure 6.
Composites of the zonal mean of temperature C_T *and specific humidity* C_Q *at the epoch for different orientations of the IMF, where* N *represents the number of cases. Solid and dashed lines indicate the 90% confidence levels. The gray region represents the longitudinally averaged Antarctic icy surface profile. Figure adapted from [57].*

leading to a vertical current $J_z = \sigma E_z$ from the ionosphere to the ground through the stratospheric and tropospheric layers.

4. The observed timescales of the response of the atmosphere to SW variations

Timescales are important in order to discriminate the involved physical processes in the SW-atmosphere coupling. In their pioneering work, Francia et al. [15] found that, during 2007 and 2008, the ULF activity, in both Pc1-2 and Pc5 frequency ranges, is correlated with the surface air temperature with different delays. Their results, shown in **Figure 7a**, indicate that the temperature is significantly correlated with the Pc1-2 power at a time lag of 1 day. Although lower, the correlation with the Pc5 power is also significant, reaching the maximum value when the temperature is delayed by 3 days with respect to the Pc5 power.

In the meanwhile, Regi et al. [57] computed the cross-correlation between high cloud cover (HCC, $h > 6$ km) over Antarctica and logPc1-2 time series at TNB during 2003–2010, band-pass filtered at ~27 days. They found several Antarctic regions characterized by significantly high correlation values. Examples of positive cross-correlations are shown in **Figure 7b**. Their experimental results suggest an average time delay of approximately 1 day of the response of the atmospheric cloud cover to the SW-driven Pc1-2 ULF power intensification. Such delay is consistent with the timescales of electrodynamics–/microphysical-related processes here proposed. In [57] a possible relationship with large-scale atmospheric transport was investigated by means of a correlation analysis between Pc1-2 power and CH_4 (~27 day band-pass filtered) concentrations from MACC dataset. CH_4 is a long life chemical tracer, commonly used to characterize the middle atmosphere transport, also analyzing it in the framework of quasi-Lagrangian or conservative coordinate systems [60] and as a proxy to validate satellite measurements [61]. The results of the correlation analysis between Pc1-2 power and CH_4, during 2003–2010 (not shown here), indicate that the correlation coefficients are generally lower and not significant and superimposable to the other tropospheric parameters. They suggest that the transport is not modulated by the 27 day periodicity of the ULF activity.

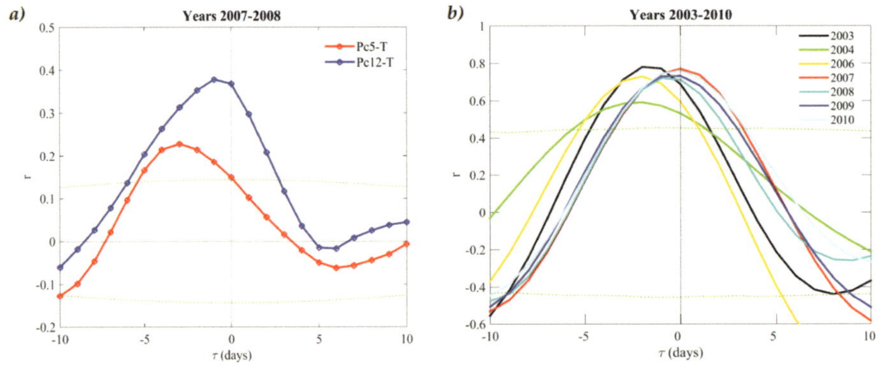

Figure 7.
(a) The cross-correlation between the Pc5 and the Pc1-2 power and the surface air temperature, at TNB, at different time lags τ. A delay τ < 0 (τ > 0) indicates that Pc1-2/Pc5 power precedes (follows) surface air temperature at TNB. The dashed green lines represent the 90% confidence level. Figure adapted from [15]. (b) Examples of positive cross-correlation analysis as a function of the delay of the HCC with respect to Pc1-2 power, observed during winter months of 2003–2010. The green lines mark the 95% confidence level. A delay τ < 0 (τ > 0) indicates that Pc1-2 power precedes (follows) HCC. Figure adapted from [57].

However, the different time delays in different years could be due to variations in large-scale transport pattern (see [57] for details).

5. Summary and conclusions

In this work we presented a short review of our experimental results regarding the possible relationship between SW and atmospheric parameters at high latitude, in Antarctica, at Terra Nova Bay. Examining possible relationship between geomagnetic activity in the Pc1-2 and Pc5 frequency range and the polar cap potential difference with stratospheric and tropospheric parameters, the results provided in [54] can be summarized as follows:

- Common power peaks emerge at ~27 days, the synodic Sun rotation period, and its first subharmonics (13.5 and 9 days), in the polar cap potential difference φ_{cap} and in the ULF activity (**Figure 3**) and in the temperature and zonal wind at tropospheric and stratospheric altitudes (**Figure 4**).

- The correspondence is more evident during winter, when solar radiation-driven processes are absent.

- Around the tropopause, approximately at 8 km, the correspondence is low, and in the stratosphere it appears mostly at the 27 day periodicity.

Further investigations [57] at tropospheric heights at high latitudes indicate that SW-driven electrodynamic processes and energetic particle precipitation related with enhancement of Pc1-2 activity can affect tropospheric temperature, specific humidity, and cloud cover. The response is quick (within ~1 day) at ground and in the troposphere. These results suggest that the electrodynamics modulate the physical properties of clouds, probably through electron scavenging microphysical mechanism. It is a matter of fact that electron scavenging is strongly dependent on vertical tropospheric-stratospheric conductivity variations, due to energetic particle precipitation driven by ULF waves, and on the vertical electric current, modulated by polar cap potential, associated to SW-magnetosphere reconnection processes. More recently, evidence of an SW signature in the mesosphere [62] has been published, indicating that the SW can affect the atmosphere through the whole atmospheric column. As discussed by [63], the processes involved in each atmospheric layer are almost certainly different, and transport phenomena could be important.

Our conclusions are supported by the observed short (<1–2 days) delay response in the atmospheric parameters at troposphere altitudes and at ground with respect to the much longer delay expected for chemical mechanism [15, 57]. However, this matter should be further investigated as underlined by [63] in particular through the examination of the time delays at stratospheric and mesospheric altitudes and at lower latitudes; the study of the dependence on different interplanetary conditions might be also useful for a more deep understanding of the atmosphere response to the SW.

Acknowledgements

The authors acknowledge J. H. King and N. Papatashvilli at NASA and CDAWeb for solar wind data (http://cdaweb.gsfc.nasa.gov) and Daniel Weimer at Space

Science Center for Space Science and Engineering Research (Virginia Tech) who provided the ionospheric electrodynamic model. The authors would also like to thank the MACC and ERA-Interim project data provider. The reanalysis data were provided by the European Centre for Medium-Range Weather Forecasts (ECMWF) and can be downloaded from the ECMWF Data Server (http://apps.ecmwf.int/datasets). Measurements of the geomagnetic field fluctuations at Terra Nova Bay are supported by the Italian PNRA (Programma Nazionale di Ricerche in Antartide, PdR2013/B2.09).

Conflict of interest

The authors declare that they are not competing interests.

Author details

Mauro Regi[1*], Marcello De Lauretis[1], Gianluca Redaelli[2] and Patrizia Francia[1]

1 Department of Physical and Chemical Sciences, University of L'Aquila, L'Aquila, Italy

2 CETEMPS, Department of Physical and Chemical Sciences, University of L'Aquila, L'Aquila, Italy

*Address all correspondence to: mauro.regi@aquila.infn.it

References

[1] de Wit TD, Watermann J. Solar forcing of the terrestrial atmosphere. Comptes Rendus Geoscience. 2010; **342**(4–5):259-272

[2] Seppälä A, Lu H, Clilverd MA, Rodger CJ. Geomagnetic activity signatures in wintertime stratosphere wind, temperature, and wave response. Journal of Geophysical Research: Atmospheres. 2013;**118**(5):2169-2183

[3] Lam MM, Tinsley BA. Solar wind-atmospheric electricity-cloud microphysics connections to weather and climate. Journal of Atmospheric and Solar-Terrestrial Physics. 2016;**149**: 277-290

[4] Rozanov E, Callis L, Schlesinger M, Yang F, Andronova N, Zubov V. Atmospheric response to NOy source due to energetic electron precipitation. Geophysical Research Letters. 2005; **32**(14):L14811

[5] Lu H, Clilverd MA, Seppälä A, Hood LL. Geomagnetic perturbations on stratospheric circulation in late winter and spring. Journal of Geophysical Research. 2008;**113**(D16):D16106

[6] Baumgaertner AJ, Seppälä A, Jöckel P, Clilverd MA. Geomagnetic activity related NO x enhancements and polar surface air temperature variability in a chemistry climate model: Modulation of the NAM index. Atmospheric Chemistry and Physics. 2011;**11**(9): 4521-4531

[7] Mironova IA, Aplin KL, Arnold F, et al. Energetic particle influence on the Earth's atmosphere. Space Science Reviews. 2015;**194**(1–4):1-96

[8] Tinsley B. The global atmospheric electric circuit and its effects on cloud microphysics. Reports on Progress in Physics. 2008;**71**(6):66801

[9] Tinsley BA, Zhou L. Initial results of a global circuit model with variable stratospheric and tropospheric aerosols. Journal of Geophysical Research. 2006; **111**(D16):D16205

[10] Tinsley BA, Burns GB, Zhou L. The role of the global electric circuit in solar and internal forcing of clouds and climate. Advances in Space Research. 2007;**40**(7):1126-1139

[11] Rycroft MJ, Nicoll KA, Aplin KL, Harrison RG. Recent advances in global electric circuit coupling between the space environment and the troposphere. Journal of Atmospheric and Solar-Terrestrial Physics. 2012;**90**:198-211

[12] Markson R. Modulation of the Earth's electric field by cosmic radiation. Nature. 1981;**291**(5813):304

[13] Simmons A. ERA-interim: New ECMWF reanalysis products from 1989 onwards. ECMWF Newsletter. 2006; **110**:25-36

[14] Inness A, Baier F, Benedetti A, et al. The MACC reanalysis: An 8 year data set of atmospheric composition. Atmospheric Chemistry and Physics. 2013;**13**:4073-4109

[15] Francia P, Regi M, De Lauretis M. Signatures of the ULF geomagnetic activity in the surface air temperature in Antarctica. Journal of Geophysical Research: Space Physics. 2015;**120**(4): 2452-2459

[16] Hartmann DL. Chapter 6-Radiative effects of clouds on earth's climate. In: Hobbs PV, editor. Aerosol-Cloud-Climate Interactions. International Geophysics. Academic Press; 1993;**54**: 151-173

[17] Intergovernmental Panel on Climate Change, editor. Climate Change

2013—The Physical Science Basis: Working Group I Contribution to the Fifth Assessment Report of the Intergovernmental Panel on Climate Change. Cambridge: Cambridge University Press; 2014

[18] Eriksson PTI, Blomberg LG, Schaefer S, Glassmeier K-H. On the excitation of ULF waves by solar wind pressure enhancements. Annales de Geophysique. 2006;**24**(11): 3161-3172

[19] Pilipenko V, Chugunova O, Engebretson M. Pc3–4 ULF waves at polar latitudes. Journal of Atmospheric and Solar-Terrestrial Physics. 2008; **70**(18):2262-2274

[20] Belakhovsky V, Pilipenko V, Murr D, Fedorov E, Kozlovsky A. Modulation of the ionosphere by Pc5 waves observed simultaneously by GPS/TEC and EISCAT. Earth, Planets and Space. 2016;**68**(1):1-13

[21] Balasis G, Papadimitriou C, Daglis IA, Pilipenko V. ULF wave power features in the topside ionosphere revealed by swarm observations. Geophysical Research Letters. 2015; **42**(17):6922-6930

[22] Regi M, De Lauretis M, Francia P. The occurrence of upstream waves in relation with the solar wind parameters: A statistical approach to estimate the size of the foreshock region. Planetary and Space Science. 2014;**90**:100-105

[23] De Lauretis M, Francia P, Regi M, Villante U, Piancatelli A. Pc3 pulsations in the polar cap and at low latitude. Journal of Geophysical Research: Space Physics. 2010;**115**(A11):A11223

[24] Lichtenberger J, Clilverd MA, Heilig B, et al. The plasmasphere during a space weather event: First results from the PLASMON project. Journal of Space Weather and Space Climate. 2013;**3**:A23

[25] Regi M, Francia P, De Lauretis M, Glassmeier KH, Villante U. Coherent transmission of upstream waves to polar latitudes through magnetotail lobes. Journal of Geophysical Research: Space Physics. 2013;**118**:6955-6963

[26] De Lauretis M, Regi M, Francia P, Marcucci MF, Amata E, Pallocchia G. Solar wind-driven Pc5 waves observed at a polar cap station and in the near cusp ionosphere. Journal of Geophysical Research: Space Physics. 2016;**121**(11): 11145-11156

[27] Glassmeier K-H, Motschmann U, Dunlop M, et al. Cluster as a wave telescope—First results from the fluxgate magnetometer. Annales de Geophysique. 2001;**19**(10/12): 1439-1447

[28] Clausen LBN, Yeoman TK, Fear RC, Behlke R, Lucek EA, Engebretson MJ. First simultaneous measurements of waves generated at the bow shock in the solar wind, the magnetosphere and on the ground. Annales Geophysicae. 2009; **27**:357-371

[29] Regi M, Del Corpo A, De Lauretis M. The use of the empirical mode decomposition for the identification of mean field aligned reference frames. Annals of Geophysics. 2016;**59**(6): G0651

[30] Francia P, De Lauretis M, Regi M. ULF fluctuations observed along the SEGMA array during very low solar wind density conditions. Planetary and Space Science. 2013;**81**:74-81

[31] Parkinson W. The influence of continents and oceans on geomagnetic variations. Geophysical Journal International. 1962;**6**(4):441-449

[32] Fujiwara S, Toh H. Geomagnetic transfer functions in Japan obtained by first order geomagnetic survey. Journal of Geomagnetism and Geoelectricity. 1996;**48**(8):1071-1101

[33] Regi M, De Lauretis M, Francia P, Lepidi S, Piancatelli A, Urbini S. The geomagnetic coast effect at two 80°S stations in Antarctica, observed in the ULF range. Annales de Geophysique. 2018;**36**(1):193-203

[34] Menk FW. Magnetospheric ULF waves: A review. In: The Dynamic Magnetosphere. Dordrecht: Springer; 2011. pp. 223-256

[35] Schulz M, Lanzerotti LJ. Particle diffusion in the radiation belts. Physics and Chemistry in Space. New York: Springer; 1974;**7**:215

[36] Elkington SR. A Review of ULF interactions with radiation belt electrons. In: Takahashi K, Chi PJ, Denton RE and Lysak RL editors. Magnetospheric ULF Waves: Synthesis and New Directions. American Geophysical Union; 2013. pp. 177-193. DOI: 10.1029/169GM12

[37] Elkington SR, Hudson MK, Wiltberger MJ, Lyon JG. MHD/ particle simulations of radiation belt dynamics. Journal of Atmospheric and Solar-Terrestrial Physics. 2002;**64**: 607-615

[38] Engebretson M, Lessard M, Bortnik J, et al. Pc1–Pc2 waves and energetic particle precipitation during and after magnetic storms: Superposed epoch analysis and case studies. Journal of Geophysical Research: Space Physics. 2008;**113**(A1):A01211

[39] Mann IR, O'Brien TP, Milling DK. Correlations between ULF wave power, solar wind speed, and relativistic electron flux in the magnetosphere: Solar cycle dependence. Journal of Atmospheric and Solar-Terrestrial Physics. 2004;**66**(2):187-198

[40] Rodger CJ, Raita T, Clilverd MA, et al. Observations of relativistic electron precipitation from the radiation belts driven by EMIC waves. Geophysical Research Letters. 2008;**35**(16):L16106

[41] Blum LW, Halford A, Millan R, et al. Observations of coincident EMIC wave activity and duskside energetic electron precipitation on 18–19 January 2013. Geophysical Research Letters. 2015;**42**(14):5727-5735

[42] Regi M, De Lauretis M, Francia P. Pc5 geomagnetic fluctuations in response to solar wind excitation and their relationship with relativistic electron fluxes in the outer radiation belt. Earth, Planets and Space. 2015;**67**:9

[43] Arnoldy RL, Engebretson MJ, Denton RE, et al. Pc 1 waves and associated unstable distributions of magnetospheric protons observed during a solar wind pressure pulse. Journal of Geophysical Research: Space Physics. 2005;**110**(A7):A07229

[44] Regi M, Marzocchetti M, Francia P, De Lauretis M. A statistical analysis of Pc1–2 waves at a near-cusp station in Antarctica. Earth, Planets and Space. 2017;**69**(1):152

[45] Mursula K, Blomberg L, Lindqvist P-A, et al. Dispersive Pc1 bursts observed by Freja. Geophysical Research Letters. 1994;**21**(17):1851-1854

[46] Dyrud L, Engebretson M, Posch J, et al. Ground observations and possible source regions of two types of Pc 1-2 micropulsations at very high latitudes. Journal of Geophysical Research: Space Physics. 1997;**102**(A12):27011-27027

[47] Rodger CJ, Clilverd MA, Green JC, Lam MM. Use of POES SEM-2 observations to examine radiation belt dynamics and energetic electron precipitation into the atmosphere. Journal of Geophysical Research: Space Physics. 2010;**115**(A4):A04202

[48] Tinsley BA, Yu F. Atmospheric ionization and clouds as links between solar activity and climate. In: Pap JM, Fox P, Frohlich C, Hudson HS, Kuhn J, McCormack J, North G, Sprigg W and Wu ST editors. Solar Variability and Its Effects on Climate. American Geophysical Union (AGU); 2013. DOI: 10.1029/141GM22

[49] Tinsley BA, Heelis RA. Correlations of atmospheric dynamics with solar activity evidence for a connection via the solar wind, atmospheric electricity, and cloud microphysics. Journal of Geophysical Research: Atmospheres. 1993;**98**(D6):10375-10384

[50] Weimer D. Improved ionospheric electrodynamic models and application to calculating Joule heating rates. Journal of Geophysical Research: Space Physics. 2005;**110**(A5):A05306

[51] Weimer D. Predicting surface geomagnetic variations using ionospheric electrodynamic models. Journal of Geophysical Research: Space Physics. 2005;**110**(A12):A12307

[52] Yagova N, Lanzerotti L, Villante U, et al. ULF Pc5-6 magnetic activity in the polar cap as observed along a geomagnetic meridian in Antarctica. Journal of Geophysical Research: Space Physics. 2002;**107**(A8)

[53] Francia P, De Lauretis M, Vellante M, Villante U, Piancatelli A. ULF geomagnetic pulsations at different latitudes in Antarctica. Annales Geophysicae. 2009;**27**(9):3621-3629

[54] Regi M, De Lauretis M, Redaelli G, Francia P. ULF geomagnetic and polar cap potential signatures in the temperature and zonal wind reanalysis data in Antarctica. Journal of Geophysical Research: Space Physics. 2016;**121**(1):286-295

[55] Torrence C, Compo GP. A practical guide to wavelet analysis. Bulletin of the American Meteorological Society. 1998; **79**(1):61-78

[56] Grinsted A, Moore JC, Jevrejeva S. Application of the cross wavelet transform and wavelet coherence to geophysical time series. Nonlinear Processes in Geophysics. 2004;**11**(5/6): 561-566

[57] Regi M, Redaelli G, Francia P, De Lauretis M. ULF geomagnetic activity effects on tropospheric temperature, specific humidity, and cloud cover in Antarctica, during 2003–2010. Journal of Geophysical Research: Atmospheres. 2017;**122**:6488-6501

[58] Laken BA, Čalogović J. Composite analysis with Monte Carlo methods: An example with cosmic rays and clouds. Journal of Space Weather and Space Climate. 2013;**3**:A29

[59] Rosenfeld D, Lohmann U, Raga GB, et al. Flood or drought: How do aerosols affect precipitation? Science. 2008; **321**(5894):1309-1313

[60] Redaelli G, Lait R, Schoeberl M, et al. UARS MLS O3 soundings compared with lidar measurements using the conservative coordinates reconstruction technique. Geophysical Research Letters. 1994;**21**:1535-1538

[61] Payan S, Camy-Peyret C, Oelhaf H, et al. Validation of version-4.61 methane and nitrous oxide observed by MIPAS. Atmospheric Chemistry and Physics. 2009;**9**(2):413-442

[62] Yi W, Reid IM, Xue X, et al. First observation of mesosphere response to the solar wind high-speed streams. Journal of Geophysical Research: Space Physics. 2017;**122**(8):9080-9088

[63] Francia P, Regi M, De Lauretis M. Solar wind signatures throughout the high-latitude atmosphere. Journal of Geophysical Research: Space Physics. 2018;**123**:4517-4520. https://doi.org/10.1029/2018JA025411

Chapter 3

Advanced HF Communications for Remote Sensors in Antarctica

Joaquim Porté, Joan Lluis Pijoan, Josep Masó, David Badia, Agustín Zaballos and Rosa Maria Alsina-Pagès

Abstract

The Antarctica is a continent mainly devoted to science with a big amount of sensors located in remote places for biological and geophysical purposes. The data from these sensors need to be sent either to the Antarctic stations or directly to the home country. For the last 15 years, La Salle has been working in the application of HF communications (3–30 MHz) with ionospheric reflection for data collection of remote sensors in Antarctica. We have developed and tested the several types of modulations, the frame structure, the radio-modem, and the antennas for two different scenarios. First, a long-range transequatorial (approximately 12,800 km) and low-power communication system is used as an alternative to satellites, which are often not visible from the poles. This distance is covered with a minimum of four hops with oblique incidence in the ionosphere. Second, a low-power system using near vertical incidence skywave (NVIS) communications provides coverage in a surface of approximately 200–250 km radius, a coverage much longer than any other systems operating in either the VHF or UHF band without the need of line of sight.

Keywords: atmosphere science, ionospheric propagation, HF, NVIS, remote sensing, advanced modulations

1. Introduction

There is a strong research activity in Antarctica in the fields of geophysics, meteorology, wildlife, flora, oceanography, and environment, among others. This research activity often involves the installation of sensors in remote places under severe/extreme weather conditions. Most of those sensors are operated with a data logger that stores the data between campaigns until it is recovered after several months.

If we need either to have access to the data throughout the year, or to install a sensor far away from the Antarctic station, we should install a radio transmission system. The standard VHF radios have a reach up to 50 km with line of sight. As the installation of a repeater station is not an option in Antarctica, it only remains to install either a satellite link or an HF link. The satellite services are expensive and are provided mostly by geostationary satellites. The geostationary orbits around the equator are not always easily visible from the poles, so the link is not fully reliable.

The HF band (3–30 MHz) is well known from the beginning of the age of the radio. The ionization of the upper layers of the atmosphere changes the direction of the radio waves in that band, so the ionosphere behaves as a mirror for a certain set of frequencies. The reflection is strongly dependent on the solar activity, the solar radiation at any time, the terrestrial magnetic field, and the angle of incidence of the wave [1]. For oblique incidence, a range up to 3000 km for a single hop can be achieved, so we can establish a link all over the planet with a few hops. The antenna in those applications should have a maximum of radiation toward the horizon.

For near vertical incidence skywave (NVIS), we can achieve a circular coverage area with a radius up to 250 km without the need of line of sight. The most suitable frequency for vertical incidence is lower than in oblique incidence, so a larger size of the antenna is needed. The antenna should have a maximum of radiation upward, so horizontal dipoles, inverted V, and loops are the best options [2]. As the distance is much lower, the transmission power required can be reduced to tens of Watt, so the application to remote sensors with power restrictions is straightforward. In the world where an increasing number of devices are going to be interconnected in an Internet of Things paradigm, a system able to communicate sensors located hundreds of kilometers away without any additional infrastructure is really welcomed.

A first step when developing a new physical layer is the sounding and characterization of the channel. Apart from the classical model of Watterson [3] for narrowband communications and Mastrangelo [4] for wideband communications, a significant amount of research has been done in ionospheric channels for a single hop at high latitudes in the Arctic [5–7]. However, few works can be found about channel sounding for long-range links with multiple hops from the Antarctica.

For the last 15 years, our research group has been working in the application of HF communications (3–30 MHz) with ionospheric reflection for data collection of remote sensors in Antarctica. In particular, we have developed a system to communicate the Spanish Antarctic Station (SAS) Juan Carlos I at Livingston Island (62.6°S, 60.4°W) and the Ebre Observatory in Roquetes (Tarragona, Spain) (40.8°N, 0.5°E). It is a 12,700 km link with multiple hops and without the use of any repeater. First, we started to sound the channel and estimate the main parameters of the channel [8–12]. Then, we have developed and tested a wide range of modulations, with its frame structure, the radio-modem, and several antennas for both the long range with oblique incidence and the near vertical incidence scenario [13–17]. We are also developing a self-organized network of NVIS nodes that can handle the delays and the unavailability of the ionospheric channel. The NVIS nodes may behave as a hub able to collect data from other neighboring sensors and transmit the joint data to the central node.

This chapter is organized as follows. In Section 2, the evolution of the hardware of the radio-modem and the antennas is presented. The modem is prepared for both channel sounding and data transmission and has evolved to low-cost software radio platforms. In Section 3, the results of more than 15 years of experience in channel sounding are presented, for both the oblique and the vertical sounding. The variability of the channel as a function of time, season, and year is summarized. In Section 4, the physical layer of the communication system of the modem is introduced. The modulation, the frame structure, and the synchronization techniques for both the long-range modem and the NVIS modem are described. The performance in terms of bit rate and bit-error rate is presented. In Section 5, new routing strategies for NVIS networks are explained. The NVIS node has to collect the data from the sensors nearby and establish a delay tolerant network (DTN) with the rest of the nodes. Finally, Section 6 contains the conclusions and some other applications of HF communications.

2. System

From the beginning of the project, we pursued a software-defined radio (SDR) hardware platform which was able to both sound the ionospheric channel and implement different modulation schemes without changing the hardware. This kind of platform was not available 15 years ago, so we had to develop our own platform with the FPGAs Virtex-IV and Spartan-II and the fastest ADC and DAC of that time. The control of the whole system and the rest of peripherals was performed by an embedded PC [10]. For the long-range link between the SAS and Spain, the system was able to transmit along the whole HF band (3–30 MHz), 24 hours a day, with two synchronized receivers connected to two wideband antennas: a monopole with antenna tuner and a wideband inverted V. The transmission power was 250 W and the system was designed to work, connected with the main power. The author is referred to [11] for further details of that system.

NVIS communication between remote sensors and the SAS is a quite different challenge. The sensor will be often battery-powered so the transmission power should be as low as possible. Although we started developing our own platform for NVIS [18], we can now find different compact SDR platforms that include FPGA, embedded microprocessor, and AD/DA converters in a very cost-effective platform. The implementation of the low-cost transceiver for remote sensors using NVIS is detailed in the following section.

2.1 The NVIS transceiver

The hardware of the system (see **Figure 1**) is composed by three parts: a Red Pitaya (RP), a Raspberry Pi 3 (Rpi3), and different peripherals.

The RP (FPGA board) is a low-cost SDR platform dedicated to the transmission and reception of the radiofrequency signals, converting them from analog to digital with an ADC of 14 bits, decreasing the frequency of the signal carrier, processing the signal, and sending the low-pass IQ components to the Rpi3 via Ethernet. Similarly, the Rpi3 can send the IQ components to the RP for the transmission. The sampling rate is 125 Ms/s at the RF plane, while it is 100 ks/s at the IQ plane.

Internally, the RP contains a Zynq® 7010, based on the Xilinx System on Chip (SoC) architecture. These products integrate a feature-rich dual-core or single-core ARM® processing system (PS) and a Xilinx programmable logic (PL) in a single

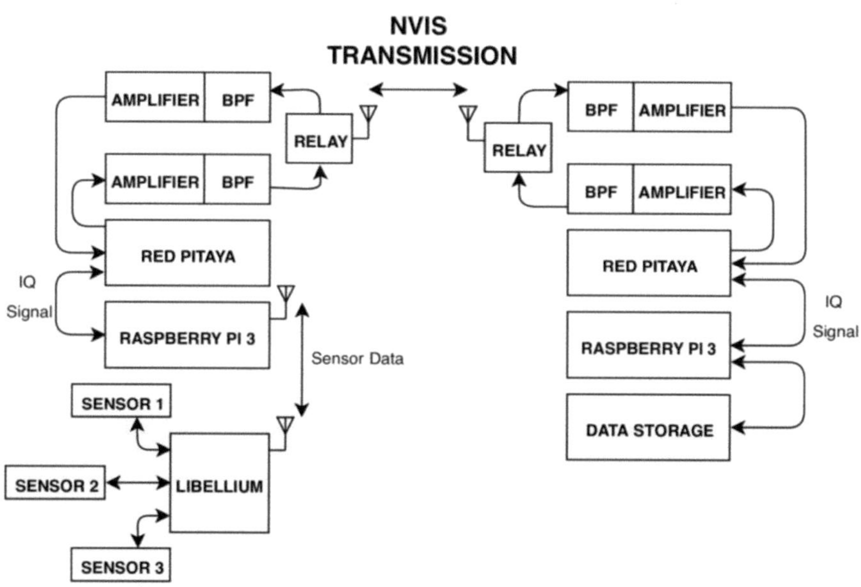

Figure 1.
Block diagram of the NVIS communication system for remote sensors.

device. On the PS, there is a Linux operating system for controlling the PL, where there are all the hardware configurations that allow the different transmission modes.

Although there is a true microprocessor inside the RP, it is not able to handle all the control functions, since the biggest part of the RP is dedicated to the real-time processing in the PL part. That is the reason for adding the Rpi3, a single-board computer that controls the overall system operation. Rpi3 is able to collect data from many wireless sensors connected via Zigbee and encapsulate them on a single frame to be sent through the NVIS channel. In our case, we have used the solution of Libelium [19], with a maximum range between 1 and 8 km, depending on the chosen solution. When Rpi3 either stores 1000 bytes of data or a defined time lapse runs out, it sends the baseband IQ components of the frame to the RP to be transited to the SAS. On the other side, the RP is waiting for reception. When the preamble of the transmission is found, the RP sends the received IQ components to the Rp3, which demodulates all the data frames and extracts the data from all the sensors.

As mentioned before, the transmission power has to be less than 10 W to ensure a proper operation in a battery-powered scenario. As the maximum output power of the DAC is 0 dBm, we need a linear minimum amplification of 40 dB. The power amplifier is controlled by an electronic circuit that switches the supply voltages of the power stages in the proper order. This hardware measures the forward and reverse power, so the Rpi3 could switch the transmission off in case of mismatch. Finally, a band pass filter (BPF) attenuates the out-of-band emissions.

At the receiver site, another BPF from 2 to 7 MHz is placed close to the antenna to filter all the unwanted noise and interferences typical of the HF band, such as AM broadcast stations. As the received signal may be around −90 dBm, an amplifier of 30 dB is needed in order to take the maximum advantage of the dynamic range of the ADC, without saturating the converter. Finally, the received data are stored in a solid-state disk.

The total measured power consumption of the system is about 7.2 W in sleep mode, which is the dominant mode. Once every hour, the transceiver transmits and receives a few seconds with a power consumption of 96 W. Under these conditions, the transceiver will operate for 2 weeks with a battery of 110 Ah.

2.2 The radiating system

When working in the HF band, we have to expect large-size antennas if we want to achieve dimensions close to half wavelength. If we are going to install the antennas in Antarctica, we will have strict environmental conditions, so no complex installations can be built. Moreover, if the antennas are for remote sensors powered by batteries, we should try to maximize the gain of the antennas in order to reduce the transmission power of the amplifier at minimum. Taking all this into consideration, the best choice are monopoles or wire antennas that need only one elevated point at most.

There is a great difference between the long range with oblique incidence and the near vertical incidence scenario. In the first case, we need a maximum of radiation toward the horizon, so a vertical monopole is the simplest option. In the second case, we need a radiation lobe with elevation angles between 70 and 90°, so the horizontal dipole or the inverted V should work better.

The monopole is the simplest antenna for long-range HF communications. It is easy to carry and install and is very robust against adverse weather conditions. As a wideband antenna, we need an antenna tuner that has to be tuned a low power before every frequency change. As the conductivity of the soil gets worse, the angle of maximum radiation is not 0° any more, but it rises up to 20–30° with respect to the horizon [10]. To prevent this, we have to install a set of radials above the soil to improve its conductivity. In the simulation, for a 7.5 m monopole over a permafrost soil, we demonstrate that 32 radials of 15 m each, we can expect a 2 dB improvement by using radials (see **Figure 2**). In **Figure 3**, we can see the monopole installed in a hill nearby the SAS and a more detailed explanation can be found in [13].

For NVIS applications, the horizontal dipole would be one of the best antennas, but it needs one mast at either end. The inverted V, however, achieves a similar performance with only one mast at the center. For the V of **Figure 4**, we have optimized the height of the antenna (Mast h), the heights of the end (Min h), and the distance from the center to the end (Yf). The length of the V changes accordingly. In **Table 1**, we can see the results for three types of soil: ideal (infinite σ, $\varepsilon_r = 6$), rural ($\sigma = 0.01$, $\varepsilon_r = 15$), and permafrost ($\sigma = 0.00005$, $\varepsilon_r = 3$), where σ and ε_r stand for the conductivity and the dielectric constant, respectively, and j is the imaginary unit. The conductivity of the permafrost causes a drop up to 5.5 dB in the gain with respect to the ideal ground. We need a mast height of 13 m to achieve a gain of 1.3 dB. As we want to have a minimum power consumption, we will operate

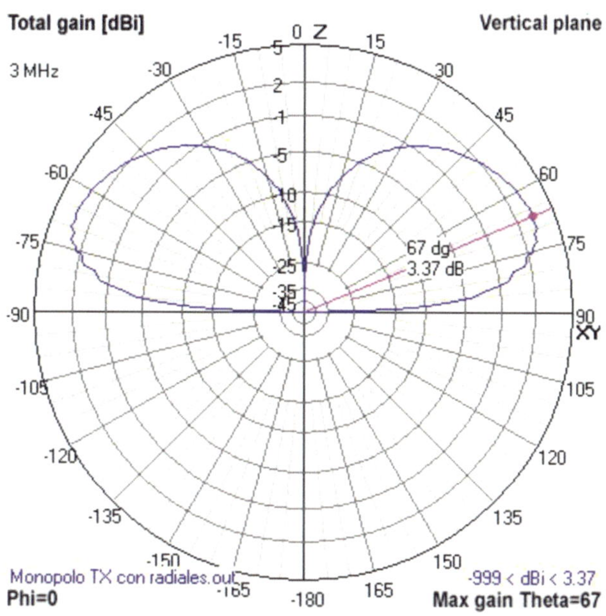

Figure 2.
Diagram of radiation of a 7.5 m monopole with radials over permafrost.

Figure 3.
Installation of the monopole with radials in the SAS.

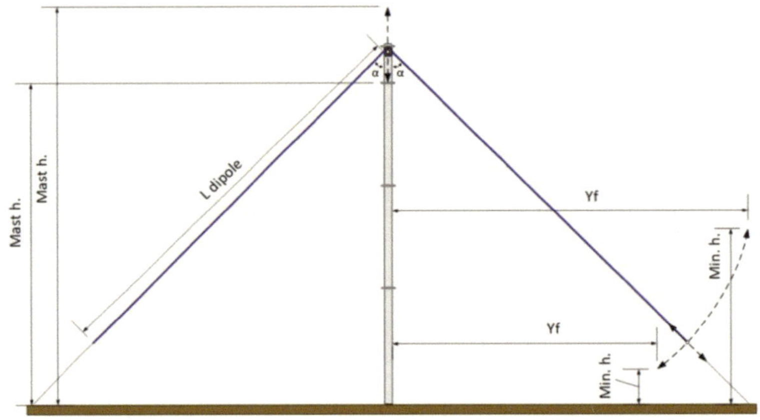

Figure 4.
Simulation parameters of the inverted V.

Soil type	Optimization algorithm	Gain (dBi)	SWR	Impedance (Ω)	Mast h. (m)	Min h. (m)	Yf (m)
Ideal	Evolve	6.8	1.96	25.6 + 3.2j	11.01	2.00	12.39
Rural	Evolve	3.8	1.05	47.7 + 0.4j	10.81	1.87	12.39
Permafrost	Evolve	1.3	1.27	63.3 – 1.0j	13.08	2.00	11.51

Table 1.
Optimization of the inverted V antenna.

at a single frequency and try to maximize the gain at that frequency. So, we do not intend to have a wideband antenna in this scenario. Checking the reports of ionograms of the last decade from Ebre Observatory [20] and Lowell Digisonde International [21], we came to the conclusion that the best frequency for maximum availability would be 4.5 MHz.

It is important to note that the optimization of the gain is a key factor for the transmission, while the radiation diagram with a maximum around 90° and a minimum at the rest of elevation angles is the most important issue in reception, because we minimize the noise and interference from the nondesired directions [22].

As far as the horizontal dipole is concerned, there is a 3 dB gain improvement, bearing in mind that two mast installations are needed. The optimum height, as discussed in [23], is between 0.16 and 0.22λ depending on the type of soil.

3. Channel sounding

The ionosphere is one of the layers of the upper atmosphere situated between about 90 and 400 km above the surface. Thanks to its atomic composition, the ionic charge allows radiofrequency signals to rebound and go back to the terrestrial surface, thus creating a communication channel. The ionization of this layer is caused by solar radiation producing the apparition of free electrons that change the refraction index of the medium. As more radiation there is, more ions are exited and the maximum reflected frequency increases. As the sun radiation is the major cause of frequency variation, daily, monthly, seasonal, and yearly variations have to be taken into account. Apart from the variation between day and night, one of the most significant variations is caused by the solar cycle reaching maximum and minimum levels approximately about every 11 years. The sunspot number (SSN) measures the number of spots visible on the face of the sun. The higher the number, the more radiation ionizes this layer of the atmosphere. **Figure 5** shows the variation of the solar cycle radiation from 1995 to 2017, and the estimation until 2019. We can point out that in the current year (2018), solar radiation is in minimum levels. This minimum level causes that the frequency used for the transmissions is much lower than in previous years (between 4.5 and 6.5 MHz during the day).

The ionosphere is observed throughout the world by a set of observatories. Most of them have an ionosonde developed by Lowell Digisonde International [21] and publish the ionogram in a common webpage [25], enabling the parameterization of the ionosphere behavior along the day in quasi-real time.

As the ionosphere is divided into several layers (D, E, F_1, and F_2), which are always moving, the ionospheric channel behaves as a slow fading multipath channel, similar to the wireless channels for mobile communications. Hence, the ionosphere can be characterized by the following parameters: time dispersion (multipath, delay spread), frequency dispersion (Doppler shift, Doppler spread), noise, interferences, and channel availability [26].

All these factors will allow us to determine the best modulations, size of the frame, and occupied bandwidth to optimize the transmission in both the long-range and the NVIS case. In the next two following points, we aim to describe the results of the sounding of both types of communications.

3.1 Oblique sounding results

When we try to characterize a channel, we have to distinguish between the narrowband analysis and the wideband analysis. The narrowband analysis allows us to

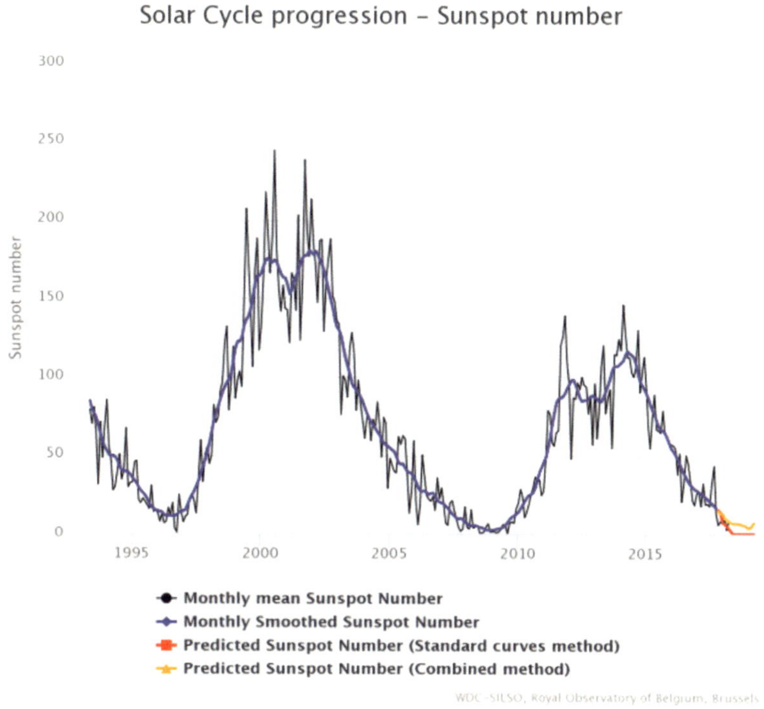

Figure 5.
Evolution of the sunspot number (SSN) [24] from 1995 to 2017.

determine the channel availability and the signal-to-noise ratio (SNR). This sounding is performed with tones that range from 2 to 30 MHz with steps of 500 Hz and are measured with a 10 Hz bandwidth during a time interval of 10 s. We concluded that a good quality of service is achieved for a SNR larger than 6 dB in a 10 Hz of bandwidth [8]. The SNR is measured after filtering the signal with some different windows (Hanning, Blackman, Flattop, Kaiser) defined in [11]. In **Figure 6**, we can see the evolution of the signal envelope during the 10 s transmission interval for each window. The SNR is measured as the ratio between the average of the signal power in case of absence of interference and the noise power measured in the seconds where there is no transmission. The Kaiser window turned out to be the best filtering technique.

The SNR and the channel availability (defined in [8]) are studied as a function of frequency and the time of day (see **Figure 7**). The big differences between day and night and between sunrise and sunset are explained in detail in [11].

The wideband analysis allows the characterization of the rest of the channel parameters such as time and frequency dispersion and wideband SNR. This analysis is performed by sending a pseudorandom noise (PN) sequence that allows us to obtain an estimation of the channel response as a function of time. The channel matrix, that is, the evolution of the channel impulse response over time, is calculated correlating the received signal with the original PN sequence. Then, the scattering function is calculated as the fast Fourier transform (FFT) of the channel matrix. A detailed analysis of the scattering function is key to determine modulation, frame length, separation between data block and data corrector, and other issues of the physical layer [9].

In **Figure 8**, we can see the averaged channel parameters for the campaign 2013–2014. During daytime, high frequencies (20–30 MHz) show the highest delay spread (up to 4 ms) and Doppler spread (up to 1.5 Hz). That means a bigger amount of intersymbolic interference and a higher degree of variability. Sunset and sunrise are the most unstable moments because the ionosphere is changing due to the ion formation or ion recombination. They are, therefore, the least suitable periods for the transmissions. Finally, nighttime is the most stable moment from 19 to 06 UTC.

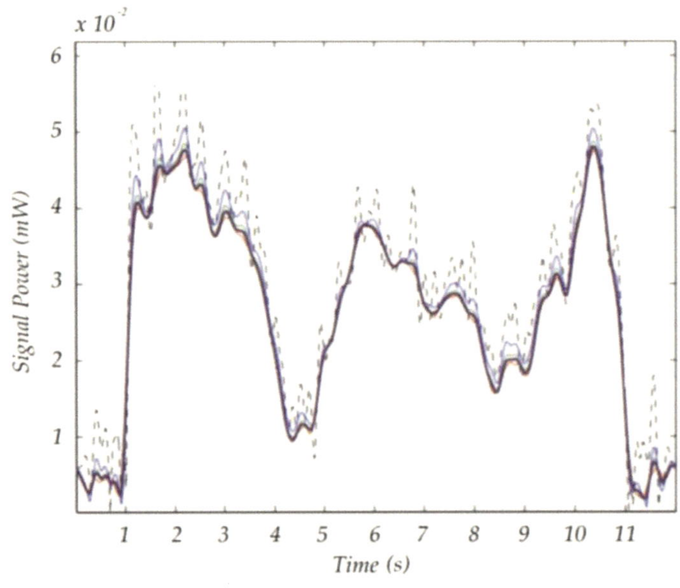

Figure 6.
Time evolution of the signal envelope for narrowband analysis.

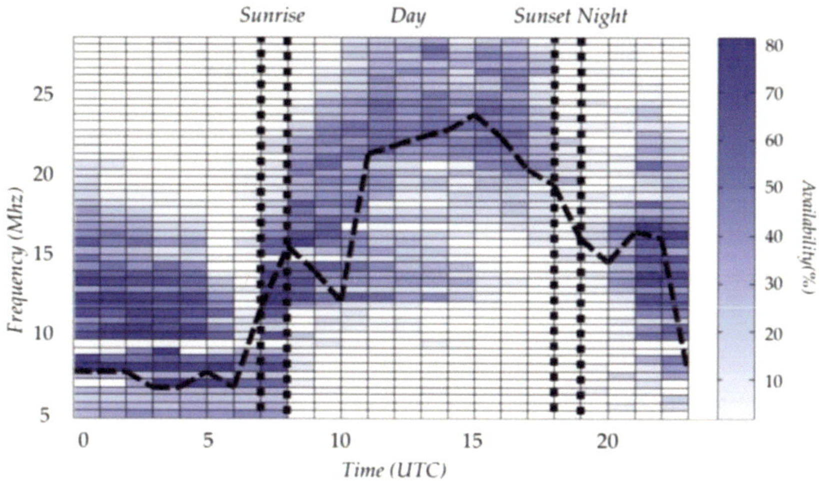

Figure 7.
SNR as a function of frequency and time of day (February 17, 2014) during the campaign 2013–2014.

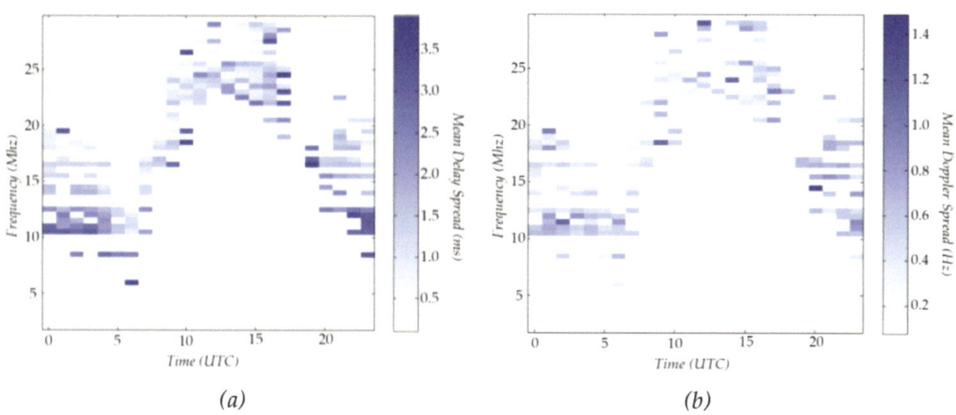

Figure 8.
Wideband channel measurements during the campaign 2013–2014: (a) mean of delay spread (in ms); (b) mean of Doppler spread (in Hz).

3.2 NVIS sounding results

For the NVIS channel, there are a few factors to be taken into account. For the narrowband analysis, we only have to check the nearby ionograms and choose the optimum working frequency as the $0.85 \times f_{oF2}$, being f_{oF2} the critical frequency of the upper layer of the ionosphere [27]. For the wideband analysis, we have to notice that the NVIS channel is the same channel that affects the ionosonde when it is measuring the height of the different layers to build the ionogram. The ionosonde temporarily stores a file with the IQ components that is used to calculate the critical frequency, virtual height, and total electron content of each layer. If we have access to this raw data file, we can have an initial estimation of the wideband channel parameters [28]. Of course, an ad hoc channel sounding will yield to better results. In **Table 2**, we can see the studies performed with the raw data of the ionosonde of Ebre Observatory. As expected, both the delay spread and the Doppler spread are lower than in the oblique sounding, so the modulation, the length of the frame, and many other parameters will allow a physical layer with higher transmission capabilities.

		October 24, 2012		March 5, 2013	
Parameters	**Wave**	**Mean**	**Variance**	**Mean**	**Variance**
Doppler spread (Hz)	Ordinary	0.681	0.094	0.378	0.199
	Extraordinary	0.123	0.095	0.081	0.061
Doppler shift (Hz)	Ordinary	−0.088	0.554	−0.025	0.208
	Extraordinary	−0.073	0.478	−0.222	0.177
Multipath spread (μs)	Ordinary	710.71	2.83	496.01	2.02
	Extraordinary	921.41	4.46	712.47	3.19

Table 2.
NVIS sounding results from data of the ionosonde at Ebre Observatory.

4. Physical layer

HF communications are often designed to operate with high-power amplifiers at the transmitter side, with typical values from 1 to 10 kW. When thinking about remote sensors supplied by batteries or solar panels, the transmitted power has to be much lower. For long-range transmissions between the Antarctic stations and Europe, a power value less than 150 W is desirable, while a value less than 30 W is expected for unattended remote sensors installed around the stations.

In a low-power scenario, the modulation has to be extremely robust with respect to noise and interference. Several replicas of the transmitted signal arrive to the receiver due to the reflection at the different layers of the ionosphere. As the layers are in constant motion, the HF channel behaves as a slow-frequency selective channel, so as the mobile channel for wireless communications. Hence, some of the latest techniques applied to the world of mobile telephony can be adapted to HF communications.

For the long-range link (about 12,760 km) with oblique incidence, the SNR at the receiver is extremely poor, often with negative values. In that situation, we defined two different modulation schemes: (i) the robust mode, for SNRs negative or close to zero and (ii) the fast mode, when the SNR is positive, and the bit rate can increase significantly.

In the robust mode, a modulation based on direct-sequence spread spectrum (DS-SS) was designed. The effective net data rate is very low (hundreds of bps), but the data can be demodulated under high levels of noise and interference. In the fast mode, the single carrier-frequency domain equalization (SC-FDE) modulation was used, since it can handle higher data rates (up to 3 kbps) if the different subcarriers can be received at a positive SNR.

For the NVIS link with vertical incidence, the situation is much more suitable. Although the transmitted power is lower, the received signal is higher while the level of interference and noise coming from the vertical direction is reduced significantly. With a SNR between 10 and 20 dB with an available bandwidth of 3 kHz, a narrowband phase shift keying (PSK) or frequency shift keying (FSK) is proposed, achieving bit rates of tens of kbps that ensure another range of applications, such as messaging, e-mail, and digital voice transmission.

When a node is working in an asynchronous mode, the receiver has to be waiting for an incoming signal continuously. A robust signal detector with low false alarm probability has to be developed, with strict requirements of energy consumption. The following subsections deal with the different modulations and the signal detector in detail.

4.1 Modulations for the long-range link

In the oblique transmission from Antarctica, two modulations have been designed for the two reception modes [13]: (i) a DS signaling [15] was designed for high robustness mode and (ii) a SC-FDE for high throughput mode [16].

4.1.1 DS signaling

The spread spectrum tests were performed using a DS-SS signaling modulation. It consists of the use of a whole family of PN sequences [26] and associating each of them to a symbol. The information about the transmission will be contained in the own sequence by means of the use of a codebook. The receiver also has the codebook available and uses it as a dictionary to obtain the transmitted information. The PN sequences used for these tests were gold sequences [29], taking advantage of their low cross-correlation in each family.

The results for the cumulative density function of the bit error rate (BER) results for two thresholds of 5 and 10% are shown in **Table 3**. They present bit rates around hundreds of bits per second—enough for data to be transmitted—and with a reasonable quality assuming that we are facing the most hostile time zone of the channel, the daytime. We found out that the best possible combination is a Gold PN sequence family 2047 length, using a bandwidth of 16.6 kHz and a final bit rate of 89 bps.

4.1.2 SC-FDE

The first tests using SC-FDE were conducted in [16], with the aim of minimizing the problems generated by the peak-to-average power ratio (PAPR) and interchannel interference (ICI) of the previous works using OFDM [17]. The single carrier-frequency domain equalization has been designed with several parameter modifications in comparison with [17], such as variable bandwidth, different length of blocks, and

PN length	Ts (ms)	BW (kHz)	Bit rate (bps)	BER (5%)	BER (10%)
2047	**123**	**16.6**	**89**	**0.73**	**0.81**
1023	123	8.3	81	0.66	0.79
511	127	4	73	0.64	0.78
255	127	2	65	0.53	0.67

Table 3.
Cumulative density function of the BER results for two thresholds of 5 and 10%. Best BER values in bold.

Constellation	Tb (ms)	Bit rate (bps)	Spectral efficiency (bps/Hz)	BER (>0.05)
PSK	10	296.3	0.74	0.54
PSK	**30**	**358.2**	**0.90**	**0.56**
PSK	**50**	**373.8**	**0.93**	**0.57**
PSK	70	381.0	0.95	0.39

Table 4.
Cumulative SC-FDE results for PSK constellation. Best results in bold.

several constellations. The reader is referred to [14] for the results of the extensive tests in SC-FDE, and a synthesis of the best results in terms of the cumulative distribution function (CDF) for a BER greater than 0.05% for a bandwidth of 400 Hz can be found in **Table 4**.

From the results presented in **Table 4**, the best possible configuration in terms of bit rate and cumulative density function of BER is the proposal with a bit time of 50 ms, in close competence with the proposal of 30 ms. The PSK using 50 ms was finally chosen because the ratio between the bit rate and the spectral efficiency against the BER was slightly higher than the same metric for 30 ms. That configuration was finally proposed to implement the high throughput mode of the oblique Antarctic transmission system.

4.2 Modulations for NVIS

The transmission system using NVIS has to be power efficient in order to attain the requirements of the battery. The tests performed are designed to optimize the required power for bit rates lower than 3 kbps, which cover the major part of remote sensing applications. The tests compare the performance of simple modulations, that is, 2-FSK, 2-PSK, 4-FSK, and 4-QAM (quadrature amplitude modulation), to find out the best modulation to be used with low transmission power, from 0.3 to 24 W. The duration for each test is adjusted so as the same amount of bits is sent for each modulation. The occupied bandwidth is 2.3 kHz to be consistent with the most common HF standards [30]. The results presented are derived from a 4-month survey between February and May 2018. The tests were performed between our premises in Barcelona (41°24′33.62″ N, 2°7′48.82″ E) and a field laboratory in Cambrils (41°4′57.22″ N, 1°4′4.61″ E) placed 96 km away. The frequency was fixed to 4.5 MHz after a detailed analysis of the ionograms from the Ebre Observatory in Roquetes, situated 80 km from Cambrils. We have only performed the transmissions during the day, so the system only uses one frequency. **Figure 9** shows the cumulative density function of the BER for a transmission power of 0.7 and 24 W. For low power transmission (0.7 W), PSK outperforms the rest of the modulations, presenting a BER less than 10^{-3} the 55% of the time and a BER less than 10^{-2} the 80% of the time. For high power (24 W), 2-FSK, 2-PSK, and 4-PSK behave in the same way, while 4-FSK has a much poorer performance. This is because FSK needs a higher bandwidth in order to keep the carrier spacing, and we have made all the tests under the same conditions.

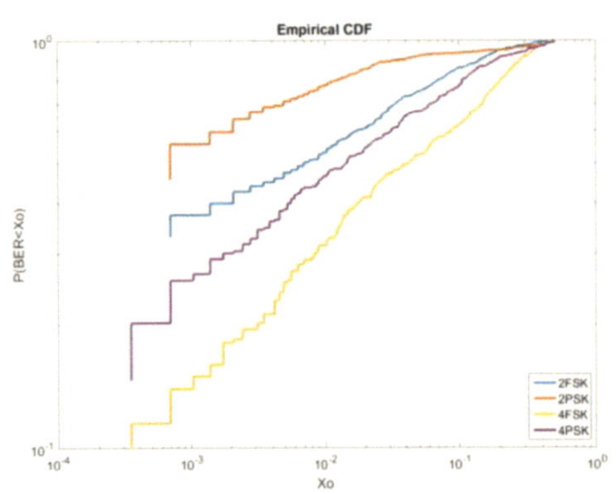

(a)

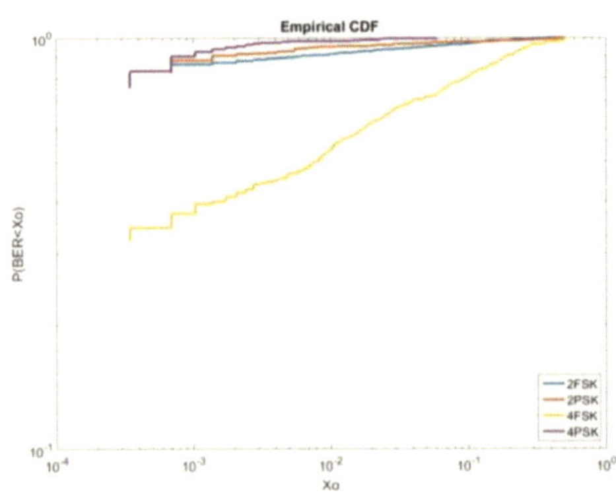

(b)

Figure 9.
CDF of the BER for an NVIS link with a TX power of 0.7 W (a) and 24 W (b).

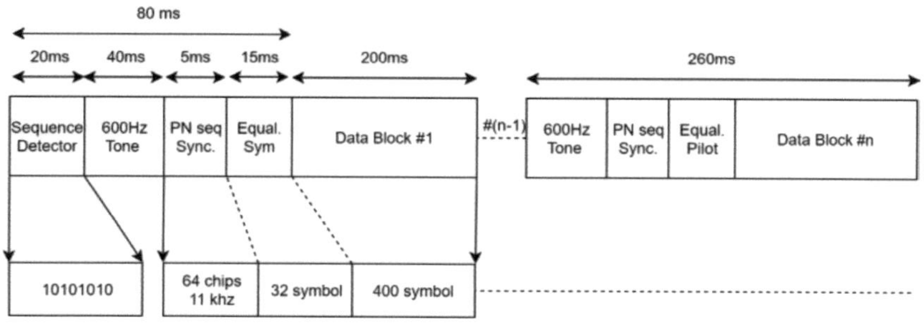

Figure 10.
Synchronization header for the NVIS frame.

4.3 Signal synchronization

In every telecommunication system, a precise time and frequency synchronization is a key issue in order to receive and demodulate the signal in the best conditions. The classical approach to time synchronization uses a PN sequence, finding the starting point of the frame by simple correlation. In practice, the clock differences between transmitter and receiver as well as the Doppler introduced by the channel may cause frequency shifts up to ±50 Hz. It follows that the received signal is rotated in phase and, therefore, hampers the correlation. An initial frequency synchronization in narrowband has to be done first. Hence, a tone of 600 Hz (with respect to the carrier) is added to detect which global frequency appears during the signal transmission. A tone of 600 Hz is often masked by the huge levels of noise and interference that are typical in the HF band. Therefore, we need a way to detect in a robust way that the signal is present with a low probability of false alarm. A known sequence of appearance of the 600 Hz tone is added at the beginning of the frame. Once the frequency shift is corrected, next step is synchronization. As the low-cost hardware is limited in speed, memory capacity, and programmable space of the FPGA, the design of the PN sequence is based on achieving correlation with the use of the smallest possible size, in the fastest way and requiring the minimum memory. A PN m-sequence of order 6 (64 chips) and 11 kHz of bandwidth was selected. The final header structure can be seen in **Figure 10**.

5. Protocols for sensor networks in Antarctica

The development of a wide area network of sensors around the Spanish Antarctic Station (or SAS) needs not only a robust physical layer, but also a robust protocol able to provide reliability, security, and tolerance to latency. In fact, we can see a remote sensor in Antarctica as a particular case of the Internet of Things paradigm. In this context, it is not wise to extend the traditional networking infrastructure based on routers to these networks for cost, efficiency, and protocol complexity reasons.

The presented work deals with the issues of utmost importance to achieve quality of service-aware (QoS-aware) communication in wireless and wired sensor networks based on standard communication protocols for the sensor networks around the SAS. The network consists of a system of distributed sensor nodes that interact among them and with infrastructure depending on applications in order to acquire, process, transfer, and provide information extracted from the physical world [31]. Those sensor nodes can be located anywhere and form an ad hoc network, which does not require a communication infrastructure. Sensor nodes are small enough to guarantee pervasiveness in the

Antarctic environment and may be able to observe a certain phenomenon, measure its physical properties, quantify this observation, and finally, transmit gathered data. Sensor nodes could also have processing and routing capabilities using either a wireless or a wired medium. In this environment, sensor networks must dynamically provide the necessary QoS depending on the type of information transmitted by sensor nodes in a multihop topology, and then, the information should be transmitted to central station through NVIS by implementing a delay tolerant network.

As the network is composed of an extensive mesh of spread nodes, they must be located in the same link layer domain to communicate among themselves. Therefore, they will use link layer mechanisms instead of network layer techniques such as IP networks or routing protocols. Consequently, communications become faster and time response turns tighter.

Each type of data may require specific requirements, for example, a critical alarm may demand strict real-time requirements while monitoring reports may not need latency requirements. In order to face these demands, network architecture must deal with several QoS profiles and it should allow discriminating and/or enforcing specific traffic differentiation.

Taking all the above into account, the ICT requirements identified for the system are as follows:

- Distributed system: the system itself is to be distributed and it must allow distributed applications.

- Simplicity: the number of protocols and APIs, and the number of different types of interfaces are kept to a minimum.

- Open system: open to other technologies (future proof) by applying existing standards whenever possible to avoid as much as possible proprietary solutions.

- High interoperability: intertechnology mesh networking between personal area network (PAN) and NVIS backhaul.

- Easy configuration: automatic neighbor discovery and plug and play capability.

- Security: security confinement (to avoid the spreading of vulnerabilities).

Due to the large scenario in which the research project is going to be deployed, different technologies will be needed in order to cover all the areas. Some technologies based on IEEE 802.15.4 are presented as wireless communication candidate technologies that work within mesh networks and they are useful for Antarctic local area network coverage. The result has to be able to support large, geographically diverse networks with minimal infrastructure, with potential millions of fixed endpoints. In the upper layers, there may be technologies such as Zigbee or 6LoWPAN.

When working at Layer2 (second layer of the open system interconnection protocol stack), the communication between two different technology domains (IEEE 802.15.4 and NVIS) involves a gateway, enabling the communication between two separate IEEE 802.15.4 domains across a NVIS domain. A Layer2 routing multihop algorithm capable of working over the obtained topology database is needed in complex network topologies. The multihop algorithm is in charge of determining the neighbors to reach a destination, and the communication with that destination will be requested from the link layer. It is important to bear in mind that the information used by the multihop algorithm can be filtered by the topology control algorithm (valid/nonvalid neighbors).

6. Conclusions and other applications

In this chapter, we have reviewed all the recent activities around the application of HF communications for the research community in Antarctica. The long-range transequatorial link aims to communicate the Antarctic station directly to the home country as an alternative to satellite communications for low bit rate applications. For a transmission power up to 250 W, two different transmission modes have been developed, the robust mode and the high throughput mode. The robust mode, which uses spread spectrum modulation, is suited for extreme channel conditions and achieves 85 bps for a bandwidth of 16 kHz for the spread signal. The high throughput mode, which uses multicarrier modulation and achieves 370 bps for a bandwidth of 400 Hz, is suited for good channel conditions. Although these bit rates are low, they are enough for most of the current sensors installed around the Antarctic stations.

The NVIS link can provide coverage in a surface of approximately 200–250 km radius without the need of line of sight. The main goal of the proposed system is to extend the influence area of the Antarctic stations with the deployment of a wide-area sensor network. When the sensors are distributed in distant zones, it is a hard work to collect the data regularly, and the data are often accessed once or twice a year. With the NVIS solution, all the researchers may get a report of the sensor data in the SAS every day, with no need of direct vision between the sensor and the SAS. The NVIS node has an internal memory that stores the data until the ionosphere conditions allow the transmission. The nodes are intended to be battery powered so the transmission power is kept to a minimum (below 10 W). For NVIS links, the bit rate ranges from 2.3 to 4.6 kbps, depending on channel conditions. On this basis, digital voice and low-resolution images can be sent apart from data from most of the sensors available on the market.

In addition to the use in the Antarctica or any other remote places, NVIS communications have a straightforward application in case of natural disasters, terrorist acts, and communications for developing countries. During a natural disaster or terrorist attack, all the conventional communication systems such as GPRS, 3G, and 4G can be seriously damaged and the communication systems will stop working properly. Our proposed NVIS system may help sanitary assistants, firefighters, police, and other emergency services to communicate during the event of a disaster. In that case, the ease in putting this system up and not needing direct vision between the nodes would be a good solution to save lives.

On the other hand, some parts of the world do not have any communication infrastructures, either because they are uninhabited areas or simply because people cannot afford the price of a conventional communication system. In places where there is no any telecom operator, the communication can only be made via HF and satellite. The NVIS system, based on a low-cost platform, allows the population of developing countries to have access to primary services, such as e-health and education.

Finally, there is a great deal of applications, which can use the proposed communication protocol architecture. They can be classified in detection (e.g., detection of temperatures exceeding a particular threshold, of unauthorized access), tracking (e.g., the tracking of workers in dangerous work environments), and monitoring (e.g., monitoring of inhospitable environments).

Author details

Joaquim Porté, Joan Lluis Pijoan*, Josep Masó, David Badia, Agustín Zaballos and Rosa Maria Alsina-Pagès
La Salle in Ramon Llull University, Barcelona, Spain

*Address all correspondence to: joanlluis.pijoan@salle.url.edu

References

[1] Davies K. Ionospheric Radio. Michael Faraday House, Six Hills Way, Stevenage SG1 2AY, UK: The Institution of Engineering and Technology, IET; 1990. Available from: http://digital-library.theiet.org/content/books/ew/pbew031e

[2] Witvliet BA, Alsina-Pagès RM. Radio communication via near vertical incidence skywave propagation: An overview. Telecommunication Systems. 2017;**66**(2):295-309. Available from: http://link.springer.com/10.1007/s11235-017-0287-2

[3] Watterson C, Juroshek J, Bensema W. Experimental confirmation of an HF channel model. IEEE Transactions on Communications. 1970;**18**(6):792-803. Available from: http://ieeexplore.ieee.org/document/1090438/

[4] Mastrangelo JF, Lemmon JJ, Vogler LE, Hoffmeyer JA, Pratt LE, Behm CJ. A new wideband high frequency channel simulation system. IEEE Transactions on Communications. 1997;**45**(1):26-34. Available from: http://ieeexplore.ieee.org/document/554283/

[5] Cannon PS, Angling MJ, Davies NC, Wilink T, Jodalen V, Jacobson B, et al. Damson HF channel characterisation-a review. In: MILCOM 2000 Proceedings 21st Century Military Communications Architectures and Technologies for Information Superiority (Cat No00CH37155). IEEE. pp. 59-64. Available from: http://ieeexplore.ieee.org/document/904913/

[6] Davies NC, Willink TJ, Angling MJ, Cannon PS. Initial Results from WHISPER; a Wideband HF Ionospheric Sounder for Propagation Environment Research. 2001. Available from: https://researchportal.bath.ac.uk/en/publications/initial-results-from-whisper-a-wideband-hf-ionospheric-sounder-fo

[7] Angling MJ, Cannon PS, Davies NC, Willink TJ, Jodalen V, Lundborg B. Measurements of Doppler and multipath spread on oblique high-latitude HF paths and their use in characterizing data modem performance. Radio Science. 1998;**33**(1):97-107. DOI: 10.1029/97RS02206

[8] Hervás M, Alsina-Pagès R, Orga F, Altadill D, Pijoan J, Badia D. Narrowband and wideband channel sounding of an antarctica to spain ionospheric radio link. Remote Sensing. 2015;**7**(9):11712-11730. Available from: http://www.mdpi.com/2072-4292/7/9/11712

[9] Ads AG, Bergadà P, Regué JR, Alsina-Pagès RM, Pijoan JL, Altadill D, et al. Vertical and oblique ionospheric soundings over the long haul HF link between Antarctica and Spain. Radio Science. 2015;**50**(9):916-930. DOI: 10.1002/2015RS005773

[10] Pijoan J, Altadill D, Torta J, Alsina-Pagès R, Marsal S, Badia D. Remote geophysical observatory in antarctica with HF data transmission: A review. Remote Sensing. 2014;**6**(8):7233-7259. Available from: http://www.mdpi.com/2072-4292/6/8/7233

[11] Ads AG, Bergadà P, Vilella C, Regué JR, Pijoan JL, Bardají R, et al. A comprehensive sounding of the ionospheric HF radio link from Antarctica to Spain. Radio Science. 2013;**48**(1):1-12. DOI: 10.1029/2012RS005074

[12] Vilella C, Miralles D, Pijoan JL. An Antarctica-to-Spain HF ionospheric radio link: Sounding results. Radio Science. 2008;**43**(4):n/a. DOI: 10.1029/2007RS003812

[13] Alsina-Pagès R, Hervás M, Orga F, Pijoan J, Badia D, Altadill D. Physical layer definition for a long-haul

HF Antarctica to Spain radio link. Remote Sensing. 2016;**8**(5):380. Available from: http://www.mdpi.com/2072-4292/8/5/380

[14] Hervás M, Alsina-Pagès RM, Pijoan JL, Salvador M, Badia D. Advanced modulation schemes for an Antarctic Long Haul HF Link. Telecommunication Systems. 2016;**62**(4):757-770. Available from: http://link.springer.com/10.1007/s11235-015-0110-x

[15] Alsina-Pagès RM, Salvador M, Hervás M, Bergadà P, Pijoan JL, Badia D. Spread spectrum high performance techniques for a long haul high frequency link. IET Communications. 2015;**9**(8):1048-1053 Available from: http://digital-library.theiet.org/content/journals/10.1049/iet-com.2014.0807

[16] Hervás M, Pijoan JL, Alsina-Pagès RM, Salvador M, Badia D. Single-carrier frequency domain equalisation proposal for very long haul HF radio links. Electronics Letters. 2014;**50**(17):1252-1254. Available from: http://digital-library.theiet.org/content/journals/10.1049/el.2014.1184

[17] Bergadà P, Alsina-Pagès RM, Pijoan JL, Salvador M, Regué JR, Badia D, et al. Digital transmission techniques for a long haul HF link: DSSS versus OFDM. Radio Science. 2014;**49**(7):518-530. DOI: 10.1002/2013RS005203

[18] Orga F, Hervas M, Alsina-Pages RM. Flexible low-cost SDR platform for HF communications: Near vertical incidence skywave preliminary results. IEEE Antennas and Propagation Magazine. 2016;**58**(6):49-56. Available from: http://ieeexplore.ieee.org/document/7605405/

[19] Libelium—Connecting Sensors to the Cloud [Internet]. [cited 2018 Jul 6]. Available from: http://www.libelium.com/

[20] Ionogrames de Livingston [Internet]. 2018. Available from: http://www.obsebre.url.edu/ca/ionogrames-de-livingston

[21] LDI, Lowell Digisonde International (Digisonde.com) [Internet]. 2018. Available from: http://www.digisonde.com/stationlist.html

[22] Witvliet BA. Near Vertical Incidence Skywave. Enschede, The Netherlands: University of Twente; 2015

[23] Witvliet BA, van Maanen E, Petersen GJ, Westenberg AJ, Bentum MJ, Slump CH, et al. Near vertical incidence skywave propagation: Elevation angles and optimum antenna height for horizontal dipole antennas. IEEE Antennas and Propagation Magazine. 2015;**57**(1):129-146. Available from: http://ieeexplore.ieee.org/document/7047674/

[24] Solar Cycle progression|Solar activity|SpaceWeatherLive.com [Internet]. 2018. Available from: https://www.spaceweatherlive.com/en/solar-activity/solar-cycle

[25] LDI, Lowell Digisonde International Station List (Digisonde.com) [Internet]. 2018. Available from: http://www.digisonde.com/stationlist.html

[26] Proakis JG. Digital communications. 4th ed. New York: McGraw-Hill; 2001

[27] Lavers C. Essential Sensing and Telecommunications for Marine Engineering Applications. 1st ed. London: Bloomsbury Publishing; 2017

[28] Hervas M, Pijoan JL, Alsina-Pagès R, Salvador M, Altadill D. Channel sounding and polarization diversity for the NVIS channel. Faro, Sweden: Nordic HF 13; Aug 2013

[29] Gold R. Optimal binary sequences for spread spectrum multiplexing (Corresp.). IEEE Transactions on

Information Theory. 1967;**13**(4):619-621. Available from: http://ieeexplore.ieee.org/document/1054048/

[30] MIL-188-110A [Internet]. 2018. Available from: http://www.wavecom.ch/content/ext/DecoderOnlineHelp/default.htm#!worddocuments/mil188110a.htm

[31] Caballero V, Vernet D, Zaballos A, Corral G. Prototyping a web-of-energy architecture for smart integration of sensor networks in smart grids domain. Sensors. 2018;**18**(2):400. Available from: http://www.mdpi.com/1424-8220/18/2/400

Chapter 4

New Perspectives on Blowing Snow in Antarctica and Implications for Ice Sheet Mass Balance

Stephen P. Palm, Yuekui Yang and Vinay Kayetha

Abstract

Blowing snow processes commonly occur over the earth's ice sheets and snow covered regions when near surface wind speed exceeds a threshold value. These processes play a key role in the sublimation and redistribution of snow, thereby influencing the surface mass balance. Prior field studies and modeling results have shown the importance of blowing snow sublimation and transport on the surface mass budget and hydrological cycle of high latitude regions. Until recently, most of our knowledge of blowing snow was obtained from field measurements or modeling. Recent advances in satellite remote sensing have enabled a more complete understanding of the nature of blowing snow. Using 12 years of satellite lidar data, climatology of blowing snow frequency has been compiled, showing the spatial and temporal distribution of blowing snow frequency over Antarctica. Other characteristics of blowing snow such as backscatter structure and profiles of temperature, relative humidity, and winds through the layer are explored. A new technique that uses direct measurements of blowing snow backscatter combined with model meteorological reanalysis fields to compute the magnitude of blowing snow sublimation and transport is also discussed.

Keywords: blowing snow, climatology, sublimation, transport, thermodynamic structure, CALIPSO

1. Introduction

Antarctica is the coldest and windiest place on earth. The ice sheet reaches a height of 4 km near the center of East Antarctica and slopes toward the coasts. The sloping terrain together with radiative cooling of the surface and lower atmosphere produces drainage flows known as katabatic winds. In some places, wind speeds can attain hurricane force and blow for days on end. It is not surprising, then, that the snow present in Antarctica is almost always airborne. Snow lifted from the surface and carried aloft by wind is known as drifting and blowing snow. Drifting snow is generally defined as airborne snow confined to a maximum height of 2 m. Once snow particles attain a height greater than 2 m, it is considered blowing snow. Blowing snow is important for a number of reasons including ice sheet mass balance [1, 2], the water budget of high latitude regions [3], and the reconstruction of paleoclimate records from the physical and chemical records obtained from ice cores [4]. Interaction with blowing snow is a major factor for changes in surface ice characteristics, such as rifts, crevasses, ridges, sastrugi, etc. and in deposition of snow on sea ice [5–7].

Most of our understanding of blowing snow comes from field measurements at the surface or from numerical modeling. Numerous observations conducted in Antarctica have measured the properties of blowing snow such as particle size, number density, mass flux, and atmospheric conditions associated with blowing snow episodes [8–10]. Most such studies are made in drifting or blowing snow conditions where the snow particles are confined to shallow layers. Particle size ranges from 50 to 450 m with the largest particles occurring near the surface. These studies also provided data on the wind speed required to initiate blowing snow. This wind speed, known as the threshold velocity, depends on the properties of the snow on the surface such as age, temperature, density, sphericity, and cohesiveness [1]. Generally, the threshold velocity ranges from about 5 to 8 m s^{-1}. All field measurements of blowing snow in the literature are made below a height of 10 m. However, blowing snow can frequently reach heights of 100 m or more and little or nothing is known about the properties of these deep blowing snow layers [11, 12]. This is mainly because when they occur, the conditions are too harsh to make measurements in the field.

Because of the lack of observations over Antarctica, the temporal and spatial frequency of blowing snow was not known until recently. With the advent of active satellite remote sensing (lidar), it was shown that blowing snow occurs more frequently than 70% of the time over large regions of Antarctica and can reach heights of 500 m [13].

2. Blowing snow detection from satellite lidar

Because of the scarcity of manned weather observation stations in Antarctica, the only practical way of obtaining information on blowing snow over the entire continent is from satellite remote sensing. While there have been passive sensors in polar orbit since the 1960s, it is very difficult to detect blowing snow from passive visible or infrared (IR) channels. In the visible, the blowing snow is indistinguishable from the underlying snow surface, and in the IR, there is generally not enough temperature contrast between the surface and blowing snow to make the latter visible. In 2003, ICESat carried the Geoscience Laser Altimeter System (GLAS) into polar orbit [14]. GLAS had both altimetry and atmospheric channels and was the first satellite lidar to study the earth's surface and atmosphere. Atmospheric data from GLAS were used to develop a technique to detect blowing snow layers from satellite lidar data. In 2006, the Cloud-Aerosol Lidar and Infrared Pathfinder Satellite Observation (CALIPSO) satellite was launched into polar orbit [15]. CALIPSO was specifically designed to study the earth's atmosphere using a multi-wavelength lidar. The data from CALIPSO have been used to study the climatology of blowing snow over Antarctica for the period 2006–2017 [16].

A lidar instrument transmits short pulses of (usually) visible or near IR laser light into the atmosphere, and a very small portion of the light is scattered directly backward from molecules, aerosols, and clouds. A telescope and associated detector electronics record the backscattered light into time-resolved "bins" usually 30–75 m in length. For satellite lidars like CALIPSO, the resulting backscatter profile extends from about 40 km in altitude to a kilometer or two below the ground. For blowing snow detection, only the backscatter data in the lowest 500 m above the surface are of interest. Since blowing snow is rooted at the ground, the algorithm first detects the large ground return signal and then interrogates the bins immediately above for elevated levels of backscatter. If the ground return is not found, then it is impossible to detect blowing snow from the backscatter profile. The lack of a ground return indicates that overlying cloud layers have attenuated the lidar beam and reduced the backscatter signal from the ground to near zero. This occurs when the overlying cloud layers have an optical depth of about 3 or greater. In general, over East Antarctica, this only happens near the coast. In the interior, cloudiness is normally quite low, and what clouds are present are usually optically thin.

When the ground return is found, the backscatter level of the lidar profile bin directly above the ground is compared to a threshold value (about 10 times the local value of the molecular backscatter). If it is greater than the threshold and the 10 m wind speed (which is on the CALIPSO data product and is obtained from the GEOS-5 global analysis product) is greater than 4 m s^{-1}, then a scattering layer has been detected. The search then continues upward for the location of the top of the scattering layer. This is defined as the bin immediately below two consecutive bins that have signal levels less than a value of about twice the local value of molecular backscatter. A few tests are made on the scattering layer in an effort to remove diamond dust from the detections. Diamond dust is fairly common in Antarctica, especially over the high Antarctic Plateau in winter and can be many km in thickness and extend to the ground. To reduce misclassification of diamond dust as blowing snow, the top of the layer cannot exceed 500 m in height (above the ground) and the backscatter level within the layer must decrease with height. If these tests are passed, then the layer is assumed to be blowing snow. Finally, the optical depth of the blowing snow layer is estimated using an extinction to backscatter a value (lidar ratio) of 25 sr that is the typical value of ice crystals found in cirrus clouds [17].

3. Characteristics of blowing snow

Blowing snow is a dynamic boundary layer phenomenon, which is only slowly revealing its many secrets. The satellite lidar measurements discussed in Section 2 have increased our knowledge significantly. **Figure 1** shows examples of blowing snow as seen by CALIPSO. Each panel of **Figure 1** is a separate CALIPSO track over East Antarctica. They show the 532-nm attenuated backscatter as measured by the Cloud-Aerosol Lidar with Orthogonal Polarization (CALIOP) lidar and represent typical wintertime blowing snow events in Antarctica. Generally, the average height of the layers is about 120 m, but they can range from just a few meters to over 400 m in depth. The lidar attenuated backscatter values range from about 1.0×10^{-5} m^{-1} sr^{-1} near the top of the layer to 5.0×10^{-4} m^{-1} sr^{-1} in the lower third of the layer. This is consistent with blowing snow models that indicate both the particle size and number density decrease with height. On average, the optical depth of blowing snow layers (not shown) is about 0.2, but can range from 0.05 to 1.0 [13]. A defining feature of blowing snow as visualized by lidar is the cellular-like structure of the backscatter. Close inspection of **Figure 1** will reveal relatively small (2–3 km), regularly spaced cell-like structures. This is a very common feature of blowing snow layers and seems to suggest that convection is occurring within the layer.

A recent discovery aided by active (lidar) and passive satellite measurements is the size of blowing snow storms in East Antarctica. Evidence shows that these storms can cover an enormous area and last for a number of days [13]. An example of such a storm is shown in **Figure 2**, which is a MODerate resolution Imaging Spectroradiometer (MODIS) false color (RGB = 2.1, 2.1, 0.85 μm) image of a large area of blowing snow covering an area about the size of Texas (695,622 km^2) in East Antarctica on October 14, 2009. This false color technique is the best way to visualize blowing snow from passive sensors. The one drawback is that sunlight is

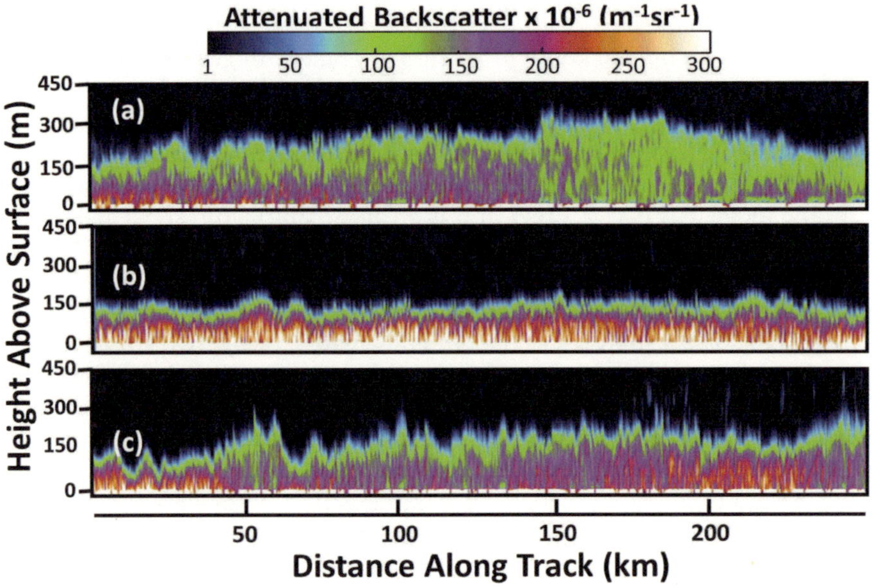

Figure 1.
CALIOP 532-nm attenuated backscatter of typical wintertime blowing snow layers over East Antarctica for (a) September 18, 2007, 09:10 UTC, (b) September 7, 2010, 14:14 UTC, and (c) October 10, 2010, 12:34 UTC.

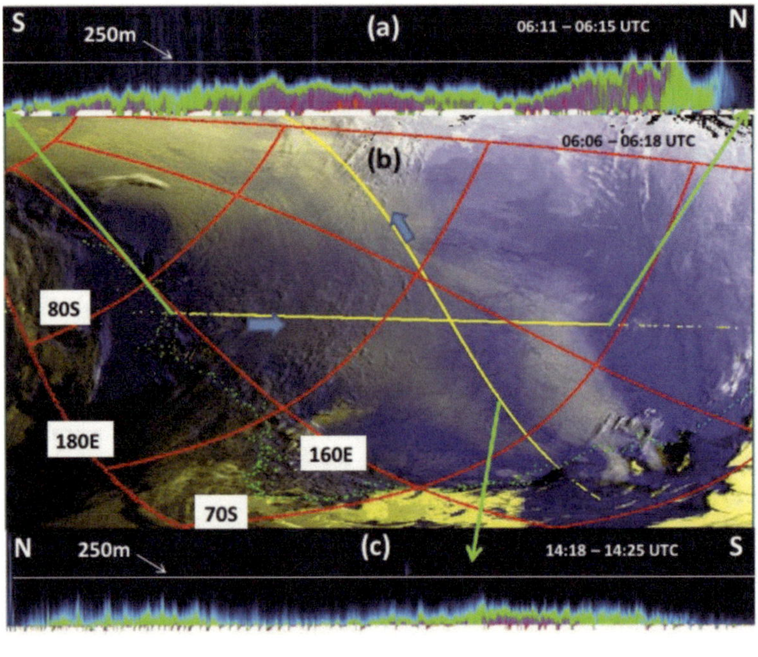

Figure 2.
A large blowing snow storm over Antarctica with blowing snow transport from continent to ocean on October 14, 2009. (a) CALIOP 532-nm attenuated backscatter along the yellow (south to north) line indicated by the green arrow as shown in (b) at 06:11–06:15 UTC. (b) MODIS false color image at 06:06:14–06:17:31 UTC showing blowing snow as dirty white areas. The coastline is indicated by the green dots, and two CALIPSO tracks, where blowing snow was detected, are indicated by the yellow lines. (c) CALIOP 532 nm attenuated backscatter along the yellow (north to south) line, 14:18–14:25 UTC.

required. In **Figure 2**, blowing snow is a dirty gray-white, the ice/snow surface (in clear areas) is blue, and clouds are generally a brighter white. Also shown in **Figure 2** are two CALIPSO tracks (yellow lines) and their associated retrieved blowing snow backscatter (upper and lower images of CALIOP backscatter). Note that the yellow track lines are drawn only where blowing snow was detected by CALIOP, and that not all the CALIOP blowing snow detections for that day are shown. The green dots denote the coastline. Plainly seen along the coast near longitude 145–150 E is blowing snow being carried off the continent. This is an important but poorly understood component of the ice sheet mass balance equation and will be discussed further in Section 5.

The temperature and moisture structure of blowing snow layers are very important to the calculation of the sublimation rate of blowing snow particles (sublimation is discussed further in Section 5). This information is typically acquired by radiosondes, but during intense blowing snow episodes, they cannot be launched due to high winds. Thus, the temperature and moisture structure of these layers remain somewhat unknown. Surface measurements during blowing snow have shown that sublimation of the snow particles will cool and moisten the air. Eventually, the air becomes saturated and sublimation ends. However, this has been shown to be true only near the surface. Measurements higher up in the blowing snow layer have not been made. Likewise, models of blowing snow also indicate that the blowing snow layer quickly saturates and sublimation reduces to near zero in a matter of hours after initiation [3, 18]. However, the models may be missing physical processes that keep the layer from reaching saturation.

Recent work has utilized dropsonde measurements to better understand the thermodynamic structure of blowing snow layers [19]. The Concordiasi Project, which occurred in the austral fall of 2010 (September 28–December 8), utilized multiple high altitude, long duration balloons to launch 648 dropsondes over Antarctica and surrounding sea ice [20–22]. The dropsondes measured temperature, moisture, and wind from the lower stratosphere to the surface. Some of the sondes fell through blowing snow layers as measured by CALIPSO and MODIS. One example is shown in **Figure 3**, which shows the CALIOP measured 532-nm attenuated backscatter in the color image with the dropsonde measured temperature (red profile) and moisture (white profile) drawn on the image. The distance between the dropsonde and the CALIPSO track was less than 5 km, but the dropsonde data were acquired about 7 hours after the CALIPSO data. However, blowing snow was very

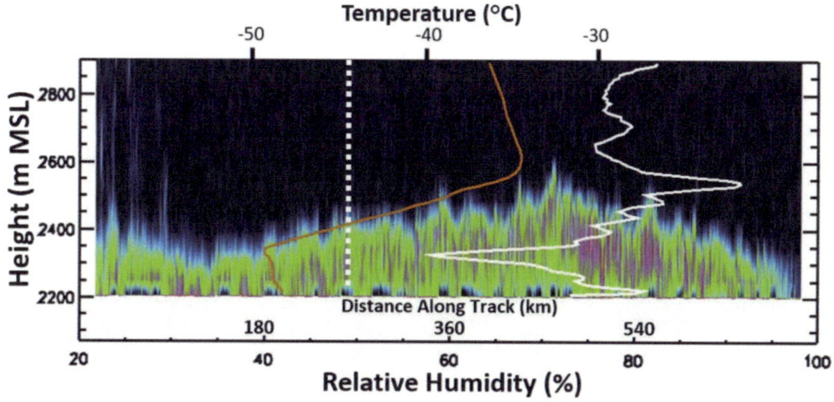

Figure 3.
CALIPSO calibrated attenuated backscatter at 05:52:50–05:54:40 UTC on October 12, 2010 showing the blowing snow layer, the approximate position of the dropsonde (vertical dashed white line), temperature (red line, scale at top), and relative humidity with respect to ice (white line, scale at bottom) as measured by the dropsonde. Note, height scale is in m above the mean sea level (MSL), and the surface is at 2200 m MSL.

likely still occurring at the time of the dropsonde (12 UTC) since CALIOP and MODIS data showed blowing snow still occurring in the area 2 hours after the dropsonde. The temperature profile in **Figures 3** and **4a** shows the normal low-level inversion beginning at about 350 m above the surface or 2550 m above mean sea level (MSL). However, it does not continue to the surface, but rather, at the height of the top of the blowing snow layer (~140–150 m above the surface), the temperature profile increases slightly as it descends. The average lapse rate in the lowest 150 m is almost adiabatic (-0.0088 versus $-0.0098°C\ m^{-1}$), which is between moist and dry adiabatic. There are even regions of the temperature profile that have a lapse rate less than dry adiabatic (**Figure 4a** between 20 and 50 m above the surface).

The relative humidity (with respect to ice) profile in **Figures 3** and **4a** shows an ample structure both above and within the blowing snow layer. Well above the blowing snow layer, the relative humidity averages about 75%. As the dropsonde descends into the blowing snow layer, and at almost the exact height of the inflection point of the temperature profile (~150 m above the surface), the relative humidity begins to increase from a value of about 60% near the top of the layer to about 82% within roughly 10 to 20 m of the surface. From there, it decreases to a value of about 75% at the surface. Most importantly, note that the relative humidity is not saturated within the layer.

The wind speed (blue line in **Figure 4b**) reaches a maximum of almost 24 m s^{-1} (~53 miles per hour) at an altitude of about 2350 m MSL, which is 150 m above the surface and very near the top of the blowing snow layer. From that altitude, the wind speed decreases linearly to roughly 15 m s^{-1} (33.5 miles per hour) close to the surface. The wind direction (red line in **Figure 4b**) varies from about 155° at 2400 m altitude to 184° at 50 m above the surface. The magnitude of the wind speed and directional shear in the lower 150 m (corresponding to the blowing snow layer) will undoubtedly produce turbulence in the layer and promote mixing. It is apparent that the mixing has destroyed the temperature inversion at the surface (assuming it existed prior to the onset of high wind speeds and blowing snow) by the process of entrainment of warmer air from above and/or adiabatic warming of the descending katabatic flow.

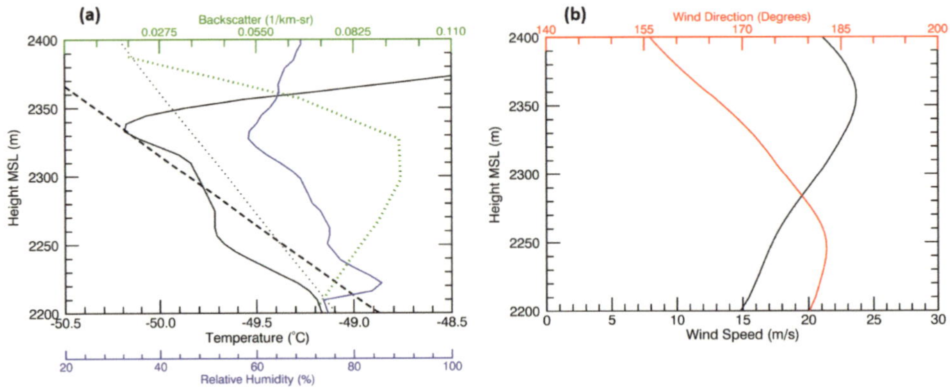

Figure 4.
(a) A magnified view of the dropsonde temperature (black solid line) and humidity (blue solid line) profiles on October 12, 2010 at 12:49 UTC and CALIPSO average backscatter (green dotted) profile at 05:50 UTC for just the lowest 200 m above the surface for the data in ***Figure 3****. Also shown is the dry (black dashed line) and moist (dotted black line) adiabatic lapse rates. (b) Dropsonde wind speed (black) and direction (red) through the blowing snow layer shown in* ***Figure 3****. Location of dropsonde: 71.61 S, 143.44 E.*

4. Climatology of blowing snow

Understanding the spatial coverage and temporal changes of blowing snow is crucial if we are to fully understand how it impacts Antarctic climate, mass balance, and hydrology. Because of the harsh climate of Antarctica and the scarcity of observations, there are few direct observations of blowing snow that cover long time periods. Most observations are near the coasts and are limited in time, covering months or a few seasons [9, 10, 23]. The CALIPSO mission has enabled the construction of a 12-year climatology of blowing snow over Antarctica. This constitutes the basis of a longer term record that can be used to examine variability and trends. The algorithm described in Section 2 has been applied to the CALIOP backscatter data acquired between 2006 and 2017. The result, shown in **Figure 5**, is the average wintertime (April–October) blowing snow frequency for that period. Note that, CALIOP began operating in June of 2006, and thus April and May are missing from that year's average. **Figure 5** indicates that large areas of Antarctica experience blowing snow more than 50% of the time. Notable patches of even higher blowing snow frequencies are found over the Megadune region east of the Transantarctic mountains, south of 75 S from 120 to 160 E, and near the Lambert Glacier along 60 to 80 E longitude. Note also that these frequency values do not contain shallow (<30 m) blowing snow layers or drifting snow, since that is the vertical resolution of the CALIOP data. In addition, blowing snow that occurs during synoptic storms is also not included as most of the time these storms contain clouds thick enough to obscure the ground (attenuate the lidar beam so that the ground cannot be detected). The latter point has a large effect only near the East Antarctic coasts and most of West Antarctica where precipitation events via synoptic storms are more frequent.

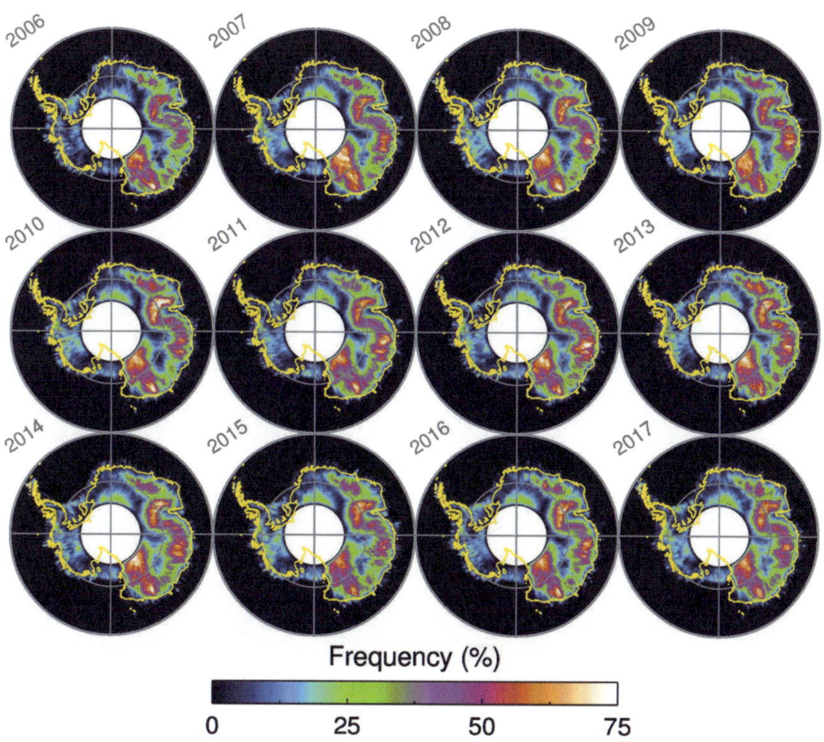

Figure 5.
The average blowing snow frequency over Antarctica for the winter months (April–October), of each year 2006–2017.

Shown in **Figure 6** is the intra-annual frequency of blowing snow for the period of 2006–2017. This is made by creating the average blowing snow frequency for each month during that period and then averaging each of the 12 months. Note that 2006 does not have data prior to June, and thus, January to May represent the average of 11 months. Blowing snow is prevalent in all months except November through February, though in these summer months, it still occurs at a reduced frequency. It is striking how fast the blowing snow frequency increases from February to March and how fast it decreases from October to November. These time periods generally coincide with the setting and rising of the sun, respectively. We hypothesize that the abrupt increase/decrease in blowing snow frequency is due to an increase/decrease in the katabatic wind flow, which is related to the increase/decrease in radiative cooling when the sun sets/rises. From April through September, large regions of Antarctica experience blowing snow more than 70% of the time with the overall spatial pattern and magnitude of blowing snow frequency remaining fairly constant through the 12 year period.

As the climate warms, it is expected that Antarctic precipitation will increase [24–27]. Climate models predict on average about a 7.4% increase in precipitation per degree of atmospheric warming [28]. Given the scarcity of observations and the problems of measuring precipitation in Antarctica, it is very difficult to verify any changes in Antarctic precipitation if indeed it is occurring. However, since blowing snow depends, at least partially, on the availability of snow [1, 29], it is reasonable to suggest that the frequency of blowing snow over Antarctica could increase as precipitation increases in a warming climate.

Using the winter average (April–October) blowing snow frequency for each year, we computed the trend of the blowing snow frequency for each 1 × 1° grid box over Antarctica for the 12 years of data. The student t-test for significance was

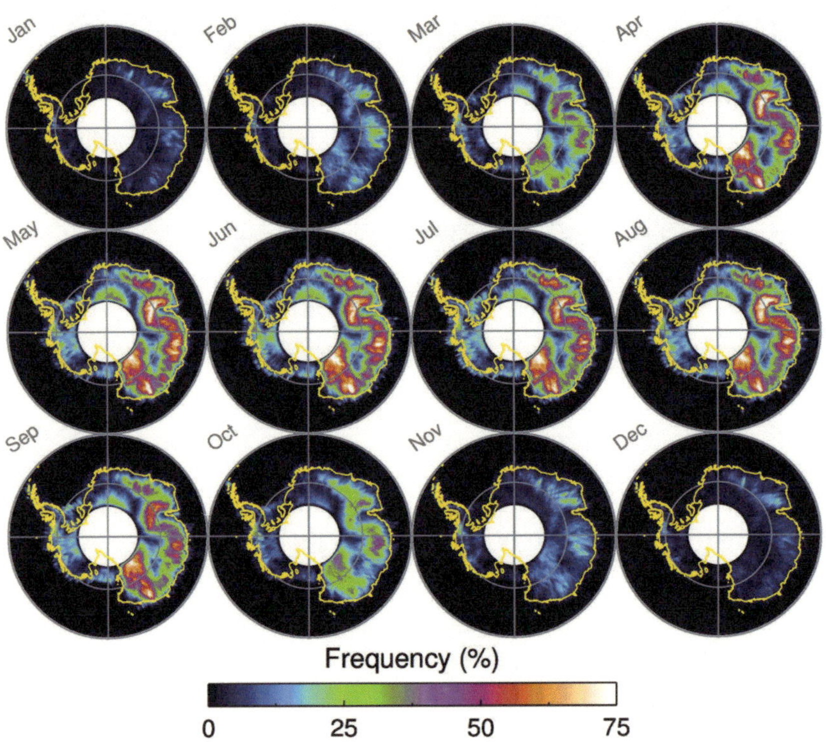

Figure 6.
The average monthly blowing snow frequency over Antarctica for the period of 2006–2017.

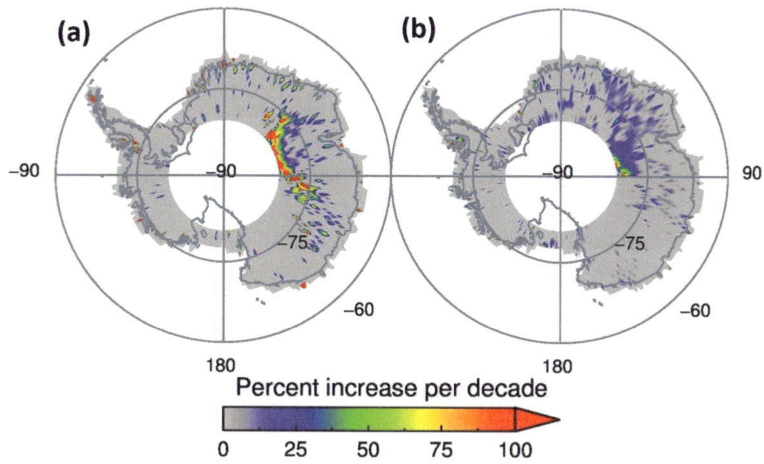

Figure 7.
Percent increase in blowing snow frequency (a) and 10 m wind speed (b) per decade for areas with trends significant at the 95% or greater level. The grid box size for the plot is 1 × 1°.

applied to each of the time series. The grid boxes that had positive trends significant at the 95% level or greater are shown in **Figure 7a**. Some very small areas of significant negative trends were also found (not shown). The color scale refers to the percentage increase in blowing snow frequency over a 10-year period. Here, we see an increase in blowing snow frequency up to 100% per decade over East Antarctica bounded between 45 and 95 E longitude. The observed geographic preference for increasing blowing snow frequency lies along the sloping edge of the ice divide. This feature in the time series trend map is intriguing. The red areas on the map in **Figure 7a** indicate grid boxes that experienced near a 100% increase in blowing snow frequency.

Since blowing snow is dependent on both the availability of snow and the near surface winds (along with the surface roughness and the snow properties), we computed the trends in 10-m wind speeds to see if the wind speed was increasing in the areas of increasing blowing snow frequency. The 10-m wind speed used was taken directly from the CALIPSO data product (V4-00), which uses the NASA Modern-Era Retrospective analysis for Research and Applications, Version 2 (MERRA-2) reanalysis. In **Figure 7b**, we see spatially similar increasing trends in 10-m wind speed in some of the areas that show increasing blowing snow frequency. However, there are a few other small areas where an increasing trend in wind speed is observed with no significant trends in blowing snow frequency.

5. Blowing snow sublimation and transport

Blowing snow sublimation and transport are two important terms in the ice sheet mass balance equation (Eq. (1)). The processes that contribute to the mass balance of a snow or ice-covered surface are precipitation (P), surface evaporation and sublimation (E), surface melt and runoff (M), blowing snow sublimation (Q_s), and snow transport (Q_t). Sublimation of snow can occur at the surface but is greatly enhanced within the atmospheric column of the blowing snow layer. The contributions of these processes to the mass balance vary greatly spatially and can be highly localized and very difficult to quantify.

$$S = \int (P - E - M - Q_t - Q_s)dt \quad (1)$$

Of the terms in this equation, precipitation is by far the greatest in magnitude followed by Q_s and Q_t. Until recently, due to the uninhabited expanse of Antarctica and the lack of observations, prior, continent-wide studies of blowing snow sublimation over Antarctica had to rely on parameterized methods that use model reanalysis of wind speed and low level moisture. The presence of blowing snow is inferred from surface temperature, wind speed, and snow age (if known). In a series of papers on the modeling of blowing snow, Dery and Yau [30, 31] develop and test a parameterization of blowing snow sublimation. Dery and Yau [2] utilize the model with the ECMWF reanalysis covering 1979–1993 and show that most blowing snow sublimation occurs along the coasts and over sea ice with maximums in some coastal areas of 150-mm snow water equivalent (swe) year^{-1}. Lenaerts et al. utilized a high resolution regional climate model (RACMO2) to simulate the surface mass balance of the Antarctic ice sheet [32]. They found drifting and blowing snow sublimation to be the most significant ablation term reaching values as high as 200 mm year^{-1} swe along the coast. There has been some recent work done on blowing snow sublimation and transport from field measurements (see for instance

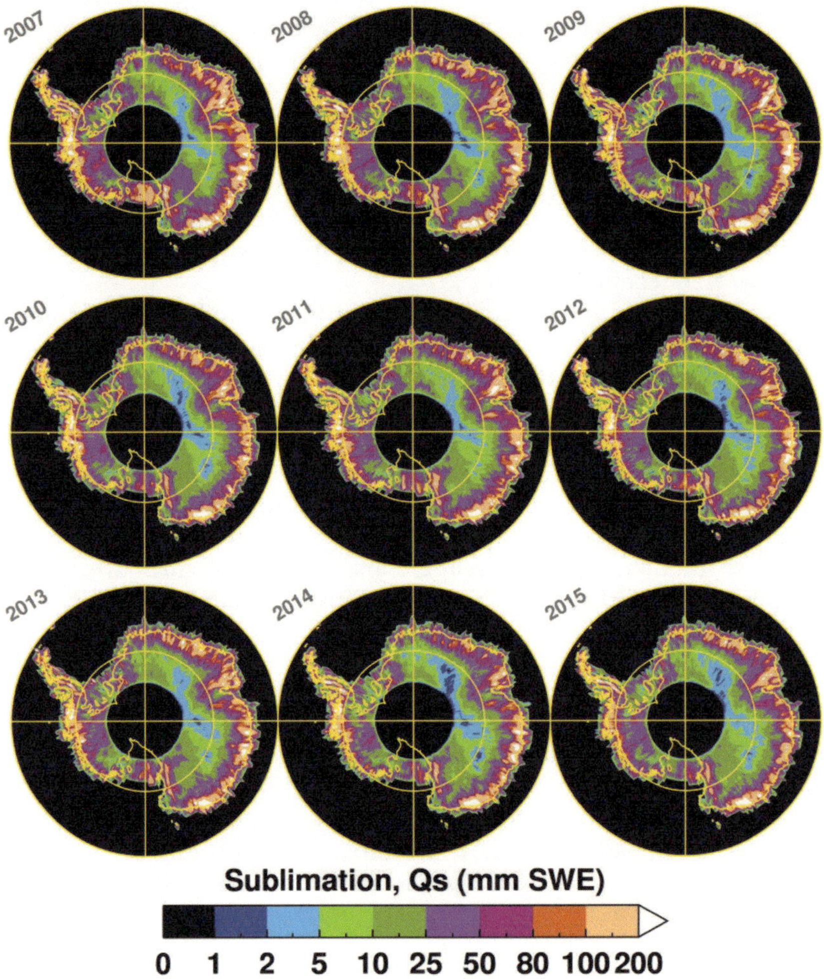

Figure 8.
Blowing snow total yearly sublimation over Antarctica for the period of 2007–2015.

[33, 34]), but the data are sparse, and the measurements are only available within the surface layer (<10 m).

While transport of blowing snow is considered less important than sublimation in terms of mass balance of the Antarctic ice sheet, erosion and transport of snow by wind can be considerable in certain regions. Das et al. [35] have shown that blue ice areas are frequently seen in Antarctica. These regions exhibit a negative mass balance as all precipitation that falls is either blown off or sublimated away. Along the coastal regions, it has been argued that considerable mass is transported off the coast via blowing snow in preferential areas dictated by topography [12]. In the Tera Nova Bay region of East Antarctica, manned surface observations show that drifting and blowing snow occurred 80% of the time in fall and winter, and cumulative snow transport was 4 orders about of magnitude higher than snow precipitation.

Considering that the accuracy of model data is questionable over Antarctica, and the complicated factors that govern the onset of blowing snow, it is difficult to assess the accuracy of model parameterizations of blowing snow sublimation and transport. However, these quantities can also be computed using direct retrieval of blowing snow layers from CALIOP attenuated backscatter and model reanalysis fields of temperature, moisture, and winds [36]. Even with this method, the

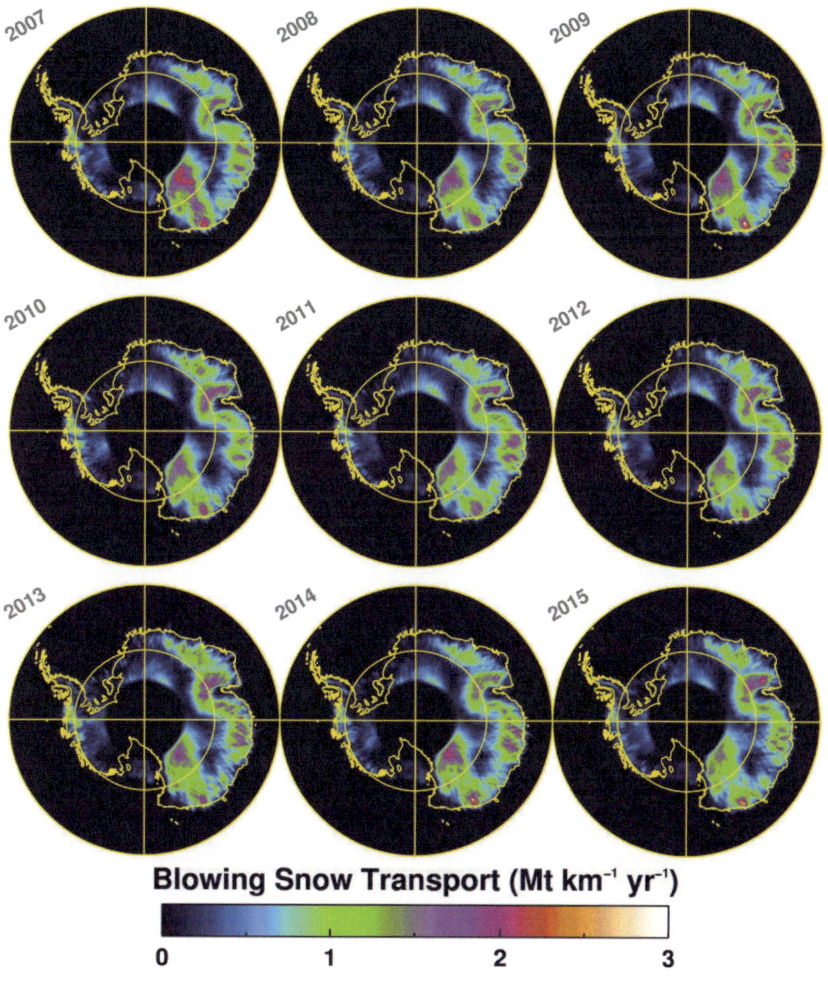

Figure 9.
The magnitude of blowing snow transport over Antarctica integrated over the year for years 2007–2015.

accuracy of the resulting sublimation is highly dependent on the accuracy of the model temperature and moisture fields. Using the CALIOP blowing snow backscatter and the MERRA-2 reanalysis, blowing snow sublimation was computed for the period of 2007–2015 (**Figure 8**). The highest values of sublimation are along and slightly inland of the coast. Notice that this is not necessarily where the highest blowing snow frequencies are located (see **Figure 5**). Sublimation is highly dependent on the air temperature and relative humidity. For a given value of the blowing snow particle density, the warmer and drier the air, the greater the sublimation. In Antarctica, it is considerably warmer along the coast but one would not necessarily conclude that it is drier there. However, other authors have noted that the katabatic winds, flowing essentially downslope, will warm and dry the air as they descend [37, 38]. Continental interior areas with very high blowing snow frequency that approaches 75% (like the Megadune region in East Antarctica) exhibit fairly low values of sublimation because it is very cold and the model relative humidity is high. **Table 1** shows the average sublimation over all grid cells in snow water equivalent and the integrated sublimation amount over the Antarctic continent (north of 82 S) for the CALIPSO period in Gt year^{-1}. Note that the 2006 data include only months June–December (CALIOP began operating in June 2006), and the 2016 data are only up through October and do not include the month of February (CALIOP was not operating). To obtain the integrated amount, the year average swe (column 1) is multiplied by the surface area of Antarctica north of 82 S and the density of ice. The average integrated value for the 9-year period 2007–2015 of 393 Gt year^{-1} is significantly greater than (about twice) the values in the literature obtained from model parameterization [39].

Transport of snow via the wind is generally important locally and does not constitute a large part of the ice sheet mass balance in Antarctica. There are areas where the wind scours away all snow that falls producing a net negative mass balance (i.e., blue ice areas), but in general, the snow is simply moved from place to place over most of the continent (**Figure 9**). At the coastline, however, this is not the case. There, persistent southerly winds can carry airborne snow off the continent. This can be seen very plainly in **Figure 2**, at the bottom right of the MODIS

Year	Average sublimation (mm swe)	Integrated sublimation (Gr year^{-1})
2006*	28.3	255
2007	56.8	514
2008	49.2	446
2009	45.3	409
2010	42.9	388
2011	47.6	431
2012	44.4	402
2013	47.7	432
2014	41.5	376
2015	41.3	374
2016*	33.2	301
Average	**43.5**	**393.4**

**2006 and 2016 consist of only 7 and 9 months of observations, respectively.*

Table 1.
The year average sublimation and the integrated sublimation over the Antarctic continent (north of 82 S) for 2006–2016.

image where snow is being blown off the coast. It turns out that this is quite a common phenomenon. Palm et al. [36] computed the mass of snow being carried off the continent by the process of blowing snow. They determined that in total about 3.68 Gt of snow is blown off the continent each year.

6. Summary and conclusion

Active remote sensing in the form of satellite lidar has given us a new perspective on, and an increased understanding of, blowing snow over Antarctica. We now know that large blowing snow storms are frequent, reach heights of 500 m, and often cover an area roughly the size of the state of Texas. From April to October, blowing snow occurs over 50% of the time over large areas of East Antarctica with some areas experiencing blowing snow 75% of the time in winter. The greatest blowing snow frequency is seen in the Megadune region of East Antarctica (south of 75 S and 120 to 160 E) and near the Lambert Glacier (60 to 80 E). Most areas of high blowing snow frequency coincide with areas of high average wind speed and/or high surface roughness. Blowing snow is prevalent in 8 months of the year, with only November through February devoid of areas of blowing snow frequency greater than 50%. Blowing snow frequency increases markedly from February to March and decreases significantly from October to November. This behavior is likely the result of katabatic wind speed increasing/decreasing as the sun sets/rises in the fall/spring.

Dropsonde and CALIOP backscatter data were utilized to investigate the temperature, moisture, and wind structure through the depth of blowing snow for the first time. The temperature structure through the layer is near isothermal, with the average lapse rate close to moist adiabatic. Above the blowing snow layer, the temperature profile is strongly stable (an inversion). The relative humidity was the greatest near the surface or slightly above (80%) and decreased through the depth of the layer with a minimum of about 60% near the layer top. Saturation was not reached within the layer indicating that sublimation of blowing snow particles was ongoing. Wind speed was 15 m s^{-1} near the surface and rapidly increased to 24 m s^{-1} near the layer top. The wind direction was constant in the lowest 50 m but backed by 25° in the upper 100 m of the layer. The near-isothermal temperature structure within the layer is likely due to the turbulent mixing of warm air from the inversion above the layer and caused by wind speed and directional shear. It is also possible that the relative humidity structure is influenced by the same process (entrainment of warmer and dryer air from above the layer), which keeps the layer from reaching saturation despite the sublimation of blowing snow particles. These results have potentially important implications for the amount of water vapor that is sublimated into the atmosphere during blowing snow episodes and also for ice sheet mass balance.

Blowing snow events identified by CALIPSO and meteorological fields from MERRA-2 were used to compute the blowing snow sublimation and transport rates. The results show that maximum sublimation occurs along and slightly inland of the coastline. This is contrary to the observed maximum blowing snow frequency, which occurs over the interior. The associated temperature and moisture reanalysis fields likely contribute to the spatial distribution of the maximum sublimation values. However, the spatial pattern of the sublimation rate over Antarctica is consistent with modeling studies and precipitation estimates. Overall, the results show that the 2006–2016 Antarctica average integrated blowing snow sublimation is about 393 $\pm$ 196 Gt $year^{-1}$, which is considerably larger than previous model-derived estimates [2, 39]. The maximum blowing snow transport amount of

5 Megatons km^{-1} $year^{-1}$ occurs over parts of East Antarctica and aligns well with the blowing snow frequency pattern. The amount of snow transported from continent to ocean was estimated to be about 3.7 Gt $year^{-1}$. These continent wide estimates of blowing snow sublimation and transport based on the direct measurements of blowing snow layers are the first of their kind and can be used to help model and constrain the surface-mass budget over Antarctica.

Acknowledgements

This research was performed under NASA contracts NNH14CK40C and NNH14CK39C. The authors would like to thank Dr. Thomas Wagner and Dr. David Considine for their support and encouragement. The CALIPSO data used in this study are the https://doi.org/10.5067/CALIOP/CALIPSO/CAL_LID_L1-STANDARD-V4-00_L1B-004.00 data product obtained from the NASA Langley Research Center Atmospheric Science Data Center. Blowing snow data are available from the author and will soon be made available from the NASA Langley Research Center Atmospheric Science Data Center. The dropsonde data were provided by NCAR/EOL under the sponsorship of the National Science Foundation and are available at https://data.eol.ucar.edu/dataset/221.002. We also acknowledge the Global Modeling and Assimilation Office (GMAO) at Goddard Space Flight Center who supplied the MERRA-2 data.

Author details

Stephen P. Palm[1*], Yuekui Yang[2] and Vinay Kayetha[1]

1 Science Systems Applications Inc., Greenbelt, Maryland, USA

2 NASA Goddard Space Flight Center, Greenbelt, Maryland, USA

*Address all correspondence to: stephen.p.palm@nasa.gov

References

[1] Gallée H, Guyomarc'h G, Brun E. Impact of snow drift on the Antarctic ice sheet surface mass balance: Possible sensitivity to snow-surface properties. Boundary-Layer Meteorology. 2001;**99**: 1-19. DOI: 10.1023/A:1018776422809

[2] Déry SJ, Yau MK. Large-scale mass balance effects of blowing snow and surface sublimation. Journal of Geophysical Research. 2002;**107**(D23): 4679. DOI: 10.1029/2001JD001251

[3] Déry SJ, Taylor PA, Xiao J. The thermodynamic effects of sublimating, blowing snow in the atmospheric boundary layer. Boundary-Layer Meteorology. 1998;**89**:251-283. DOI: 10.1023/A:1001712111718

[4] King JC, Anderson PS, Vaughan DG, Mann GW, Mobbs SD. Wind-borne redistribution of snow across an Antarctic ice rise. Journal of Geophysical Research. 2004;**109**: D11104. DOI: 10.1029/2003JD004361

[5] Déry SJ, Tremblay LB. Modeling the effects of wind redistribution on the snow mass budget of polar sea ice. Journal of Physical Oceanography. 2004;**34**:258-271. DOI: 10.1175/ 1520-0485(2004)034<0258: MTEOWR>2.0.CO;2

[6] Leonard KC, Tremblay LB, MacAyealand DR, Jacobs SS. Interactions of wind-transported snow with a rift in the Ross Ice Shelf, Antarctica. Geophysical Research Letters. 2008;**35**:L05501. DOI: 10.1029/ 2007GL033005

[7] Leonard KC, Maksym T. The importance of wind-blown snow redistribution to snow accumulation on Bellinghausen sea ice. Annals of Glaciology. 2011;**52**(57):271-278

[8] Walden VP, Warren SG, Tuttle E. Atmospheric ice crystals over the Antarctic Plateau in winter. Journal of Applied Meteorology and Climatology. 2003;**42**:1391-1405

[9] Mann GW, Anderson PS, Mobbs SD. Profile measurements of blowing snow at Halley, Antarctica. Journal of Geophysical Research-Atmospheres. 2000;**105**(D19):24491-24508. DOI: 10.1029/2000JD900247

[10] Nishimura K, Nemoto N. Blowing snow at Mizuho station, Antarctica. Philosophical Transactions of the Royal Society A. 2005;**363**:1647-1662. DOI: 10.1098/rsta.2005.1599

[11] Mahesh A, Eager R, Campbell JR, Spinhirne JD. Observations of blowing snow at the South Pole. Journal of Geophysical Research. 2003;**108**(D22): 4707. DOI: 10.1029/2002JD003327

[12] Scarchilli C, Frezzotti MM, Grigioni P, De Silvestri L, Agnoletto L, Dolci S. Extraordinary blowing snow transport events in East Antarctica. Climate Dynamics. 2010;**34**:1195-1206. DOI: 10.1007/s00382-009-0601-0

[13] Palm SP, Yang Y, Spinhirne JD, Marshak A. Satellite remote sensing of blowing snow properties over Antarctica. Journal of Geophysical Research. 2011;**116**:D16123. DOI: 10.1029/2011JD015828

[14] Spinhirne JD, Palm SP, Hart WD, Hlavka DL, Welton EJ. Cloud and aerosol measurements from GLAS: Overview and initial results. Geophysical Research Letters. 2005;**32**: L22S03. DOI: 10.1029/2005GL023507

[15] Winker DM, Vaughan MA, Omar A, Hu YX, Powell KA, Liu ZY, et al. Overview of the CALIPSO mission and CALIOP data processing algorithms. Journal of Atmospheric and Oceanic Technology. 2009;**26**:2310-2323. DOI: 10.1175/2009jtecha1281.1

[16] Palm SP, Kayetha V, Yang Y. Toward a satellite-derived climatology of blowing snow over Antarctica. Journal of Geophysical Research: Atmospheres. 2018;**123**. https://doi.org/10.1029/2018JD028632

[17] Chen WN, Chiang CW, Nee JB. Lidar ratio and depolarization ratio for cirrus clouds. Applied Optics. 2002;**41**: 6470-6476. DOI: 10.1364/Ao.41.006470

[18] Bintanja R. Snowdrift suspension and atmospheric turbulence part II: Results of model simulations. Boundary-Layer Meteorology. 2000;**95**:369-395

[19] Palm SP, Yang Y, Kayetha V, Nicolas JP. Insight into the thermodynamic structure of blowing snow layers in antarctica from dropsonde and CALIPSO measurements. Journal of Applied Meteorology and Climatology. 2018 (in review)

[20] Rabier F et al. The Concordiasi field experiment over Antarctica: First results from innovative atmospheric measurements. Bulletin of the American Meteorological Society. 2012;**94**: ES17-ES20. DOI: 10.1175/BAMS-D-12-00005.1

[21] Rabier F et al. The Concordiasi project in Antarctica. Bulletin of the American Meteorological Society. 2010; **91**(1):69-86. DOI: 10.1175/2009BAMS2764.1

[22] Boylan P, Wang J, Cohn SA, Hultberg T, August T. Identification and intercomparison of surface-based inversions over Antarctica from IASI, ERA-Interim, and Concordiasi dropsonde data. Journal of Geophysical Research—Atmospheres. 2016;**121**: 9089-9104. DOI: 10.1002/2015JD024724

[23] Takahashi S. Characteristics of drifting snow at Mizuho station, Antarctica. Annals of Glaciology. 1985; **6**:71-75. DOI: 10.3189/1985AoG6-1-71-75

[24] Agosta C, Favier V, Genthon C, Gallée H, Krinner G, Lenaerts JTM, et al. A 40-year accumulation dataset forAdelie Land, Antarctica and its application for model validation. Climate Dynamics. 2012;***38***(1–2):75-86. DOI: 10.1007/s00382-011-1103-4

[25] Frieler K, Clark PU, He F, Buizert C, Reese R, Ligtenberg SRM, et al. Consistent evidence of increasing Antarctic accumulation with warming. Nature Climate Change. 2015;**5**:348-352. DOI: 10.1038/nclimate2574

[26] Krinner G, Magand O, Simmonds I, Genthon C, Dufresne J-L. Simulated Antarctic precipitation and surface mass balance at the end of twentieth and twenty-first centuries. Climate Dynamics. 2007;**28**(2–3):215-230. DOI: 10.1007/s00382-006-0177-x

[27] Ligtenberg SRM, van de Berg WJ, van den Broeke MR, Rae JGL, van Meijgaard E. Future surface mass balance of the Antarctic ice sheet and its influence on sea level change, simulated by a regional atmospheric climate model. Climate Dynamics. 2013;**41** (3–4):867-884. DOI: 10.1007/s00382-013-1749-1

[28] Palerme C, Claud C, Dufour A, Genthon C, Wood NB, L'Ecuyer T. Evaluation of Antarctic snowfall in global meteorological reanalyses. Atmospheric Research. 2017;**190**: 104-112. DOI: 10.1016/j.atmosres.2017.02.015

[29] Gossart A, Souverijns N, Gorodetskaya IV, Lhermitte S, Lenaerts JTM, Schween JH, et al. Blowing snow detection from ground-based ceilometers: Application to East Antarctica. The Cryosphere. 2017;**11**: 2755-2772. DOI: 10.5194/tc-11-2755-2017

[30] Dery SJ, Yau MK. Simulation of blowing snow in the canadian arctic using a double-moment model. Boundary-Layer Meteorology. 2001;**99**: 297-316

[31] Dery SJ, Yau MK. A bulk blowing snowmodel. Boundary-Layer Meteorology. 1999;**93**:237-251

[32] Lenaerts JTM, van den Broeke MR, Dery SJ, van Meijgaard E, van de Berg WJ, Palm SP, et al. Modeling drifting snow in Antarctica with a regional climate model: 1. Methods and model evaluation. Journal of Geophysical Research-Atmospheres. 2012;**117**:D05108. DOI: 10.1029/2011jd016145

[33] Barral H, Genthon C, Trouvilliez A, Brun C, Amory C. Blowing snow in coastal Adélie Land, Antarctica: Three atmospheric-moisture issues. The Cryosphere. 2014;**8**:1905-1919. Available from: www.the-cryosphere.net/8/1905/2014/. DOI: 10.5194/tc-8-1905-2014

[34] Trouvilliez A, Naaim F, Genthon C, Piard L, Favier V, Bellot H, et al. Blowing snow observation in Antarctica: A review including a new observation system in Adélie Land. Cold Regions Science and Technology. 2014. DOI: 10.1016/j.coldregions.2014.09.005

[35] Das I, Bell RE, Scambos TA, Wolovick M, Creyts TT, Studinger M, et al. Influence of persistent wind scour on the surface mass balance of Antarctica. Nature Geoscience. 2013;**6**:367-371. DOI: 10.1038/Ngeo1766

[36] Palm SP, Kayetha V, Yang Y, Pauly R. Blowing snow sublimation and transport over Antarctica from 11 years of CALIPSO observations. The Cryosphere. 2017;**11**:2555-2569. DOI: 10.5194/tc-11-2555-2017

[37] Gallée H. A simulation of blowing snow over the Antarctic ice sheet. Annals of Glaciology. 1998;**26**:203-205

[38] Bintanja R. Modelling snowdrift sublimation and its effect on the moisture budget of the atmospheric boundary layer. Tellus. 2001;**53A**:215-232

[39] Lenaerts JTM, van den Broeke MR, van de Berg WJ, van Meijgaard E, Munneke PK. A new, high-resolution surface mass balance map of Antarctica (1979–2010) based on regional atmospheric climate modeling. Geophysical Research Letters. 2012;**39**:1-5. DOI: 10.1029/2011gl050713

Chapter 5

Water Balance and Thermal Regime of Lakes in Antarctic Oases

Elena Shevnina, Ekaterina Kourzeneva and Mohammad Nuruzzama

Abstract

The chapter aims to revise the capabilities of a water balance modelling approach to be applied on climate-related or practical studies of lakes located in specific conditions of Antarctica. The seasonal water balance equation (WBaL) of a lake was suggested for the lakes located in the vicinity of the Antarctic scientific stations: Bellinshausen, Progress and Maitri. First, the methods and models used to evaluate the income and outcome terms of the WBaL from minimal observational datasets are considered. Then the historical observations available on the lakes Kitezh, Priyadarshini, Stepped, Nella, Progress and Reid are described based on the technical reports of the Finnish, Indian and Russian Antarctic research programmes and from open source publications. Finally, practical recommendations on improving temporal hydrological network are formulated to give a simple solution for the seasonal water balance studies of the Lake Priyadarshini.

Keywords: Antarctic oasis, lakes, freshwater resource, climate change, water quality and quantity, human activity, water supply

1. Introduction

Ice-free areas named oases provide a home for numerous lakes in Antarctica. The size of the lakes vary from small, shallow ponds to big, deep natural reservoirs; however, these lakes are very sensitive to changes of the physical parameters of the near-surface atmosphere. The precipitation and air temperature are among the climate variables affecting the liquid water storage of the Antarctic lakes. The water balance of lakes is an important indicator of changes in climate, and it includes various sets of freshwater inflow and outflow terms depending on the time scale of the physical processes considered. On a seasonal scale, an increase of freshwater outflow from the Antarctic continent leads to changes of ocean water salinity and can affect the Southern Ocean circulation [1]. At the same time, an increase of air temperature leads to an increase in the ratio of evaporation to precipitation for a lake surface, which reduces the number of small ponds in the Antarctic oases [2]. In turn, the drying lakes may accelerate local warming in the oases. Recently, the decrease in the number and volume of small lakes is mentioned by [2–4].

While a majority of studies discuss the observed and expected changes in an air temperature regime and ice and snow cover evolution on the continent [5], the

IntechOpen

knowledge of surface hydrology and freshwater resources is still limited to the lakes in Antarctic oases [6]. This is due to the remoteness of the continent, which results in the high cost of the hydrological observational network. In this situation, a modelling approach provides the only opportunity to study regional specifics of the Antarctic lakes.

The water balance equation of a lake (WBaL) is among the traditional models used to study a quantity of freshwater accumulated in lakes [7]. The equation provides a relationship between inflow and outflow terms (precipitation, evaporation, surface inflow and outflow, etc.) and includes different components depending on lake type, a human factor or a time scale of physical processes to be accounted for. The WBaL also allows for the optimisation of the total amount of water withdrawal for human needs. This task has become interesting for the study of lakes located in the vicinity of scientific stations. In this chapter, the applicability of the WBaL to evaluate the seasonal freshwater cycle for the lakes was considered.

Models need observations to be calibrated, validated and corrected. In Antarctica, the atmospheric forcing data available for evaluations of the WBaL terms are usually limited to the standard meteorological observations within the World Meteorological Organisation (WMO) network, and temporary hydrological observations operate during summer field campaigns. The temporary hydrological observations are supported by national Antarctic programmes, and the content of data may vary year to year. Since only observations with consistent measurement techniques and instruments can be used for climate-related studies, it is important to have uniform programmes to study the water balance components of lakes in Antarctica. Then, specifics of the polar environment as well as features of logistic operations can be accounted for in solutions of the network to monitor the lake's water balance components. In this chapter, the historical hydrological observations are analysed from the technical reports of the three national Antarctic programmes and open source publications.

This chapter aims to revise the capability of the water balance modelling approach to be applied in climate-related or practical studies of lakes located in specific conditions of Antarctica. The seasonal water balance equation was suggested for the lakes located in the vicinity of the Antarctic scientific stations Bellinshausen, Progress and Maitri. First, the methods and models used to evaluate the income and outcome terms of the WBaL from minimal observational datasets are considered. Then, the historical observations available on the lakes Kitezh, Priyadarshini, Stepped, Nella, Progress and Reid are described based on the technical report of the Finnish, Indian and Russian Antarctic research programmes and previous publications. Finally, practical recommendations for the temporal hydrological network for the water balance studies of Lake Priyadarshini are formulated as an example of a simple solution.

2. Lakes in the Antarctic oases

This chapter presents an overview of the observations available to evaluate the seasonal water balance for the lakes located in the vicinity of the three Antarctic bases. Kitezh Lake is located on the Fildes Peninsula, on King George Island, and serves the water supply of the Russian Bellinshausen and the Chilean Presidente Eduardo Frei Montalva stations year-round. The Fildes Peninsula is an ice-free area, and only the northeast edge is covered by Collins Glacier. The surrounding landscape is formed by hills; the bedrock is composed of volcanic deposits mostly of basalts, tuffs and andesites [8, 9].

Lake Priyadarshini is situated on the Schirmacher Oasis, Dronning Maud Land, and it provides water for the Indian Maitri station with about 25 people on a winter crew. During the summer, the number of people doubles. The Schirmacher Oasis is located approximately 80 km from the sea coast and has the ice-free area elongated in a narrow strip around 17 km long and 3 km wide located between two glaciers. The relief of the oasis is typically hill rocks and moraines with local depressions occupied by more than a hundred lakes of different types: small, shallow ponds, big, deep lakes and shallow lakes [10, 11].

Lakes Stepped, Nella, Progress and Reid are among more than 150 surface freshwater lakes located in the Larsemann Hills, Princess Elizabeth Land [12, 13]. Lake Stepped provides the technical water supply of the Russian Progress station, which has year-round operations with 25 people, and gives the additional inflow to the water supply system of the Chinese Zhongshan station. Drinking water is extracted from the Lake Progress located within several kilometres of both stations, and it is delivered by vehicles once a week during the summer. Lake Nella is located about half as far from the stations as Lake Progress, and its water resources may be an alternative water supply to both stations. In case of an increase in activities on the Australian Low/Racowita Antarctic station, Lake Nella can become a good source of water for this station as well.

2.1 Models

2.1.1 Water balance equation of a lake

The water balance equation of a closed lake (WBaL) generally includes two terms [7]:

$$\Delta V = I - O \tag{1}$$

where ΔV is the volume change per time unit and I and O are the input and output terms correspondingly. The input and output terms consist of a number of water balance components depending on a time interval of changing in a volume. In fact, the shorter time interval requires a higher number of water balance components to be accounted for due to the inclusion of physical processes of smaller scales. On seasonal scales, the input terms of WBaL are precipitation and surface groundwater inflow runoff, while the output terms are evaporation and surface/groundwater outflow runoff. The number of terms in the WBaL also depend on the lake type, for instance, the only output term for endorheic lakes is evaporation. This type of lake is among the most common in the Antarctic oases. The human factor should also be accounted for among the output terms of the WBaL for the lakes serving the water supply for Antarctic stations.

Figure 1 shows the seasonal terms of WBaL for lakes that are used or may potentially be used as the water supply of the three Antarctic stations. The WBaL can be written as follows:

$$\Delta V = P - E \pm Y \pm G \pm W \tag{2}$$

where V is the volume of a lake, P is the precipitation (input term), Y is the surface inlet/outlet runoff (input/output terms), E is the evaporation (output term), G is the sub-surface inflow/outflow runoff (input/output terms) and W is a term connected to the human activity (a pure water consumption or a wastewater utilisation). Two terms, P and E, are related to the lake surface.

To finalise the WBaL, each term should be estimated from specific observations or modelled from standard observations. In Antarctica, it is a challenge to organise

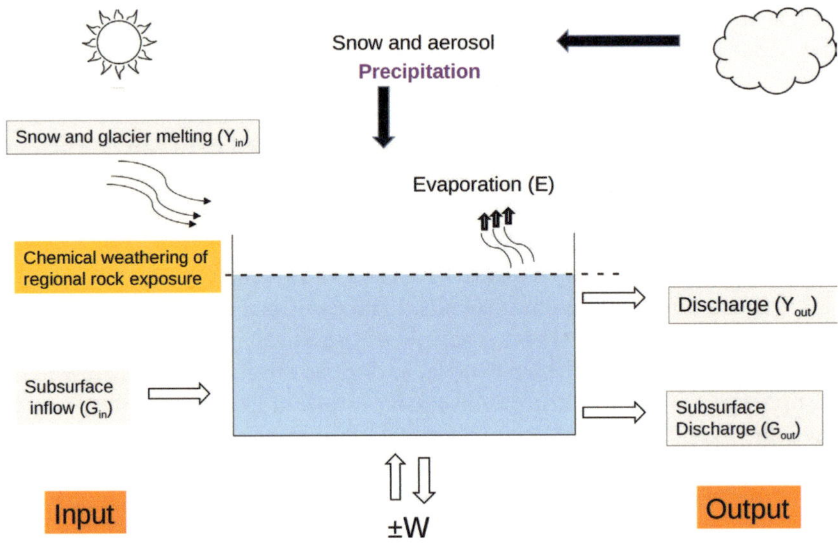

Figure 1.
The components included in the water balance equation of the lakes.

the regular observational network for monitoring the terms of water balance due to high supporting costs; thus, the modelling approach is only a way to study climate-related changes in lakes located on Antarctic oases. Modern hydrology provides numerous models to evaluate terms of the WBaL, such as surface/sub-surface inflow/outflow runoff or evaporation; however, the models still need measurements for testing experiments.

2.1.2 WBaL: the input terms

For the lakes studied, to estimate the seasonal precipitation amount, the direct measurements are available at the nearest meteorological station operating with the World Meteorological Organisation (WMO). The stations use standard methods and instruments to measure the precipitation as well as the air temperature, relative humidity, wind, solar radiation, soil temperature, etc. These meteorological observations are available for a long-term period; thus, they serve as a good source for the forcing variables to model the WBaL terms, and the inflow/outflow (both surface and sub-surface) runoffs are among the other known sources.

Melting water from a seasonal snow cover produces a surface stream flow network to transport water, sediments and dissolved minerals into a lake during a summer season. To simulate the surface/sub-surface runoff from standard meteorological observations, physically based distributed hydrological models are common [14]. These models require datasets on climatology as well as detailed soil type maps on height resolution, which do not exist for Antarctica. A spatial dimension of the local surface runoff network and the size of watersheds are too small to be described by the physically based distributed hydrological models. To use this type of models, the special geodesic measurements and geological studies are needed to reproduce the details of the local digital elevation model (DEM) and soil types [15].

The lumped hydrological models provide an alternative for the physically based models to evaluate surface runoff in climate-related and practical studies. In this case, the surface runoff is also simulated from the standard meteorological variables, and the simulations are not required on the detailed DEM or soil type data. However, the direct measurement of the river water discharges is needed to calibrate and verify the models. In seasonal water balance calculations, the sub-surface runoff inflow term can be neglected, as it was assumed in this study. The precipitation and surface inflow terms are connected to biogeochemical processes going on in lakes, as they are chief contributors of the water input to the lakes and also provide a reaction medium for terrestrial weathering. These processes ultimately affect the lake water biogeochemistry.

2.1.3 WBaL: the output terms

The evaporation from a lake surface term remains among the poorly studied variables for the lakes located in the Polar Regions. The direct measurements of evaporation are difficult to organise even in low latitudes since techniques usually require very specific instruments. Indirectly, the Dalton-type empirical equations are usually used to calculate the evaporation from standard meteorological observation [16]. In hydrology, the daily evaporation rate (mm/day) is calculated as follows:

$$\mathrm{E} = 0.14(e_0 - e_{200})(1 + 0.72w_{200}) \quad (3)$$

where e_0, hPa is the water vapour pressure at the saturation point; e_{200}, hPa is the screen level water vapour pressure and w_{200}, ms^{-1} is the wind speed. The equation is obtained using measurements with the “pan evaporation technique” on lakes located in the north of Russia [16]. Thus, this equation may give noised results in the estimation of the evaporation rate [4].

The evaporation term provides a connection between the thermal and water balance of a lake. Since the evaporation is connected with the latent heat flux from the lake surface, it can also be calculated using the lake heat balance and thermodynamic lake models. In this case, a thermodynamic lake model is forced by standard meteorological observations, namely by the air temperature, specific humidity and the wind speed. Downward short-wave and long-wave radiation, if not measured, may be estimated from cloudiness observations, using astronomical formula and empirical methods [17, 18]. Evaporation, as well as other turbulent fluxes, may then be calculated by an atmospheric surface layer block of a lake model from the simulated lake surface temperature and atmospheric variables. The example of the calculation for the evaporation in Antarctica using the thermodynamic lake model may be found in [4]. In this study, the lake model FLake [19] is used to simulate lake water temperature profiles and the turbulent fluxes from a lake surface, including the evaporation, for the lakes in the Larsemann Hills and the Fildes Peninsula. The results are sensitive to the parameters of lakes, namely, to the lake depth and the lake water transparency.

The surface outflow term can be estimated only from direct observations, since this term of the WBaL depends on both morphological features of a lake and on meteorological variables. The water withdrawal/releasing form a human factor could be estimated only from direct measurements. In water management, these two terms allow regulation of a lake’s storage in an accumulation/releasing water resources depending on human needs.

2.1.4 Volume of a lake

A bathymetrical map of a lake is among the tools to evaluate the basic morphometrical characteristics of a lake including the width, the depth (average and maximum), the surface area and the volume of water storage. To obtain the bathymetry, the space-distributed measurements on lake depth are needed. Presently, the estimates on the length, the width, the surface area and depth are published for a number of lakes on the Antarctic oases [3, 4, 7, 8, 12, 20]; however, there are still minimal data

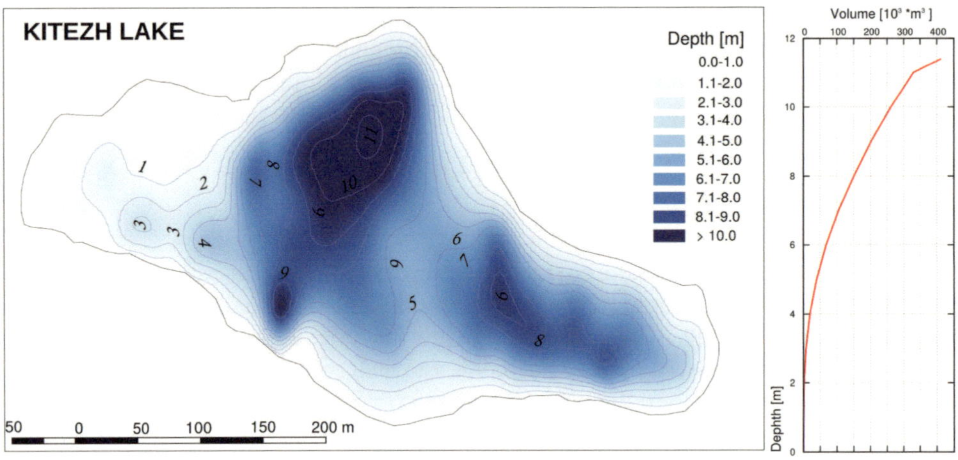

Figure 2.
The bathymetry (left) and volumetric curve (right) of Lake Kitezh [23].

on the volume of lake storage. The volume of a lake can be also evaluated from the high-resolution digital elevation model as suggested in [21]. However, the uncertainties of such estimations with freely available DEMs are still unclear [22].

To finalise the water balance of a lake, the sum of the terms should be equal to the volume changes during the selected time step (day, decade, month, season, etc.). For that, the measurements on the water level-named stage (H, m) are required together with a lake's volumetric curve (see **Figure 2** for the example). The volumetric curve is a unique lake attribute; however, there are also theoretical models allowing evaluation of lake bathymetry from the remote sensing data [22].

2.2 Observations

In modern times, the water balance components of a lake are measured occasionally, and only during the summer season (December–March) on a number of lakes located in the vicinity of the Antarctic bases. However, the content on the hydrological datasets vary year by year as well as spot by spot and in terms of a number of lakes, the location of hydrological gauges and instruments used for measurements. In this part of the chapter, we analysed the technical reports on the filed campaigns 2011–2018 published by the Finnish, Indian and Russian national Antarctic research programmes as well as more early publications in an attempt to understand whether we have measurements to support climate studies on the seasonal water balance of the lakes used for human water supply for the selected spots in Antarctica.

2.2.1 WBaL: the input terms

For Lake Priyadarshini, the measurements on the surface runoff inflow/outflow as well as snow cover measurements were not founded or do not exist. Generally, the hydrological observations on Lake Priyadarshini usually consist of the water temperature profiles and chemistry [10, 11]. A similar situation was founded for Lake Kitezh, where the measurements of water balance components are limited by one season year, 2011–2012 [23]. The temporal hydrological network on the watershed of Lake Kitezh included discharge gauges on inlet/outlet rivulets as well as snow cover extent, thickness and density. The preliminary modelling results of the surface inflow suggest that one-/two-degree "black-box" hydrological models show a good fit to the observations of the discharges on the inlet gauges.

The hydrological observations on the input terms of the WBaL for the lakes Stepped, Nella, Progress and Reid (the Larsemann Hills) usually cover the period of 1–2 months. During these periods, the temporal gauges to observe the water stage/discharge are placed in the inlet rivulets of all of these lakes. The daily river discharges are usually estimated from the water stage/discharge curve.

The seasonal observations of snow cover properties (snow depth, extent and density) are available for Lake Stepped for four summer seasons (2011–2012; 2013–2014; 2013–2014 and 2016–2017), and the preliminary results suggest that surface/sub-surface water inflow due to seasonal melting of snow contributes more than 5–7% of the total volume of storage [24–26]. The seasonal surface inflow due to snow melting is also estimated for the lakes Nella and Reid; however, the detailed study is still needed to model the surface inflow runoff term of the WBaL of these lakes.

2.2.2 WBaL: the output terms

The evaporation from open surface of Lake Priyadarshini measured during the period of January 2018 [2] with the method of eddy covariance [27]. The preliminary results of the campaign suggest that the daily values of evaporation rate ranged from 0.9 to 1.6 mm/day and strongly depend on wind speed, air temperature and lake water temperature. The results of this field experiment will be presented in a separate study in cooperation with Miguel Potes and Daniela Franz. The direct observations of evaporation were not founded on other lakes, including Kitezh, Stepped, Progress, Nella and Reid.

2.2.3 Volume and water level

The bathymetry of lakes Kitezh, Priyadarshini, Stepped, Progress and Reid are evaluated from GPS and depth measurements [4, 11, 23]. The volume of Lake Nella ($103.3\ m^3 \times 10^4$) is estimated after [21]; however, the bathymetric map of Lake Nella has not been published or ever done. Only two datasets on the Nella Lake depths were found; however, the results are only on the lake depth profiles [12, 24].

Table 1 gives the length of the observational period, the date and the resolution of the measurements of the lake water level for the period 2011–2018. The water level (stage) is measured on the lakes selected; however, the temporal hydrological network contains a different number of the observational gauges.

The measurements on the water level/stage are available daily or once every 3 days for the summer seasons 2012–2014 and 2017 in the Larsemann Hills. These data look promising to be applied in the seasonal water balance calculations and the simulation of water balance terms. However, the temporal hydrological networks on the lakes Stepped, Progress, Nella and Reid are relocated yearly, and the datasets on water level (stage) observations have a different "zero level". This "zero-level" issue makes an inconsistence in the year by year datasets for the water level/stage observations [23, 24, 28]. This problem is solved during the field campaign 2016–2017 [26].

The survey of the technical reports and publications shows that the hydrological observations of the water balance components of the lakes in the Antarctic oases occur occasionally and cover mostly only the summer season. The measurements are often inconsistent in terms of gauges' location, instruments used and measurement techniques applied. This issue is inherent in the measurements on the lakes that are provided a water supply. In this context, some harmonisation of the measurement methods and tools is needed to support climate-related study of a water balance of lakes in the Antarctic oases. Further, the hydrological observational network will be discussed, for example, on Lake Priyadarshini (the Schirmacher Oasis).

Spot	Period	Number of the gauges/measurement points or profiles/samples					
		E	SR	LST	SC	GR	CS
FA	05.01.2012–05.04.2012	0	6	12	3	0	25
SH	28.12.2017–08.02.2018	1	1	3	0	0	3
LA	27.12.2011–05.02.2012	0	3	9	1	0	48
	21.12.2012–28.01.2013	0	4	10	1	4	25
	25.12.2013–03.03.2014	0	2	6	1	0	38
	05.01.2016–20.02.2017	0	0	5	4	0	67

Note: E, the evaporation; SR, surface runoff; LST, the lake stage and water temperature; SC, snow cover; GR, sub-surface runoff and CS, the chemistry sampling.

Table 1.
The summary for the direct measurements of the water balance components on the lakes located in the Fildes Peninsula (FA), the Larsemann Hills (LA) and Schirmacher (SH) oases.

2.3 Optimisation of the temporal hydrological network

The hydrological measurement campaign to study the water balance components of Lake Priyadarshini was carried out during summer season 2016–2017. The campaign was mostly focused on evaporation [2], while the additional hydrological observations were organised on the temporal hydrological network on the watershed of Lake Priyadarshini (**Figure 3**).

The water level and temperature gauge named Parwati was equipped with the temperature and water level sensor HOBO by Onset. The evaporation gauge named Irgason was equipped with the Integrated CO_2 and H_2O Open-Path Gas Analyser and 3-D Sonic Anemometer by the Campbell Scientific and temperature sensor iButton by Maxel. The water discharges were only measured episodically on the gauge named West Inflow with the micro-current metre (GR-55) by the Gidrometpribor for reasons to be discussed in a separate paper. However, continued discharge measurements are needed to apply the modelling approach to estimate the water balance component of Lake Priyadarshini.

Figure 3 provides an example of a simple solution for the location of the temporal hydrological gauges on the watershed of the Antarctic lakes; however, the network must also include the snow measurement profiles, groundwater level gauges, etc. The good examples of the temporal hydrological network on the watershed of Lake Stepped are given in [23–26].

2.3.1 WBaL: the input terms

The input terms of the WBaL of Lake Priyadarshini are formed by the surface/sub-surface inflow runoff and wastewater that come from the Maitri station. To evaluate the surface/sub-surface runoff, a simple hydrological "black-box" model can be applied; however it should be calibrated against measurements. In this context, the field measurements of the snow cover and soil properties in the watershed of Lake Priyadarshini are important in the modelling of water balance income terms. It should be noted that for the seasonal water balance, calculations on the field measurements are needed since remote sensing observations still remain coarse for such calculations for watersheds of a small size. In addition to snow and soil measurements, the gauges named West/East Inflow and Outflow gauges should be equipped with water temperature, level and discharge sensors. The water input

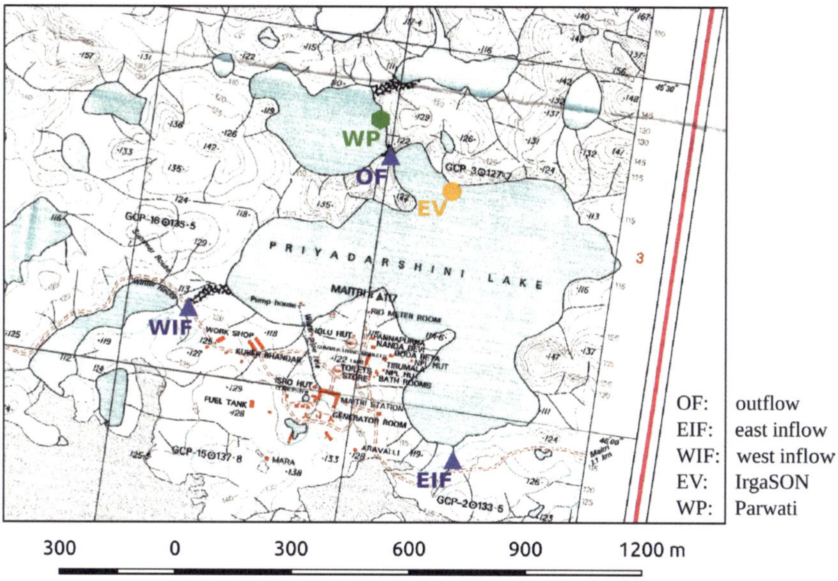

Figure 3.
The temporal hydrological network on the watershed of Lake Priyadarshini (from the field campaign of 2017–2018): the green hexagon is the water level and temperature gauge; the blue triangles are the discharge gauges on inlet/outlet streams, and the yellow circle is the evaporation gauge.

to Lake Priyadarshini due to human activity would be recommended to measure the outlet of the water purification system of the Maitri station.

2.3.2 WBaL: the output terms

The output terms of the WBaL of Lake Priyadarshini are the evaporation, surface/sub-surface outflow and water withdrawal for the needs of the Maitri station. The evaporation can be calculated from thermal balance of the lakes or from Dalton-type equations using the standard meteorological observations. The FLake model [19] simulates the evaporation rate from the open lake surface; however, this model is not yet tested against direct measurements of evaporation. The experiment on the Irgason evaporation gauge 2017–2018 (**Figure 4**) was designed to measure the evaporation rate to be used in modelling experiments with the FLake model. The preliminary results show that the daily values of the evaporation rate ranged from 0.9 to 1.6 mm/day and strongly depend on wind speed, air temperature and lake water temperature [2]. At the same time, the observational period for the campaign was too short, which may contribute limitations to the model testing results. Thus, the experiment would be interesting to continue further. The surface/sub-surface outflow should be measured as mentioned above. The water withdrawal should be measured on the inlet of the water pumping station of the Maitri station.

2.3.3 Volume and water level

The detailed study of the volume of Lake Priyadarshini is reported by [11] about 20 years ago; thus, the volume of the lake may differ. The actual GPS survey and the lake depth measurements may contribute to the actual state of the lake bedrock. The water level/stage gauge with temperature sensor (or profiled sensors) and water level data logger should be deployed on Lake Priyadarshini for the year-round operation.

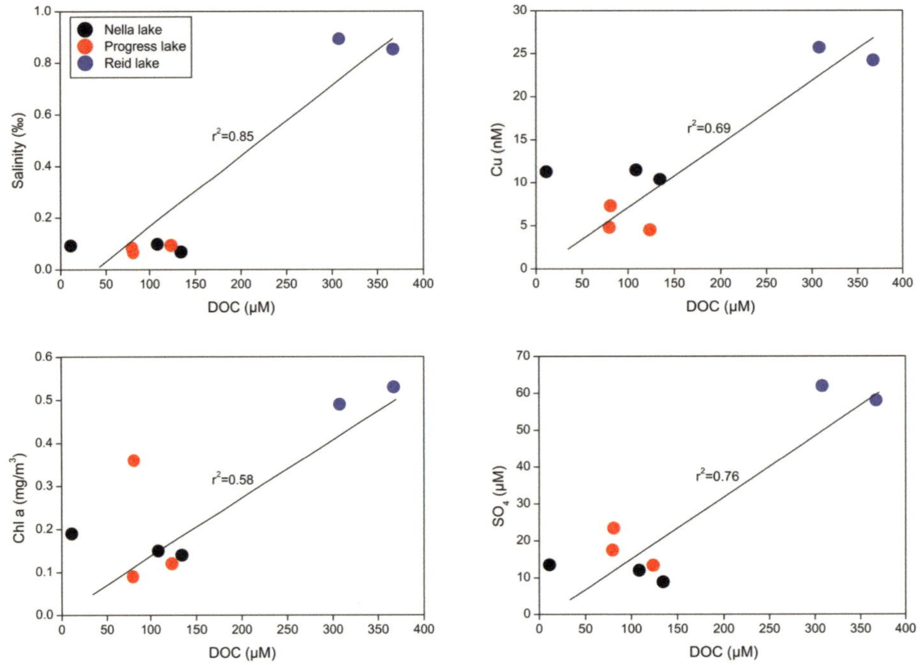

Figure 4.
Correlations between salinity, chlorophyll (Chl-a), dissolved sulphate (SO_4) and dissolved copper (Cu) and dissolved organic carbon (DOC) for water samples collected during January and February 2017 from the lakes of Larsemann Hills, East Antarctica.

2.4 Antarctic lakes: archive for terrestrial and biogeochemical processes

There is meagre knowledge on a well-developed river system in Antarctica unlike in the tropical and temperate regions; however, the Larsemann Hills regions of the East Antarctic peninsula, an ice-free oasis, has almost 150 surface freshwater lakes [12, 13]. Lakes in the region are exposed with highly metamorphosed granitic gneiss and a thin layer of sediment patches [29]. Chemical and physical weathering of the exposed lithology is pathways for nutrient supply to the lake water primary produces. Among them, silicate weathering is one of the major processes which control the atmospheric CO_2 draw down and in turn governs the long-term climatic changes [30–32]. Since then, Antarctica has been assumed to be pristine; these lakes can be a suitable archive to study the terrestrial and biogeochemical processes.

2.4.1 Sampling and analysis

To understand the terrestrial and biogeochemical processes in the region, Antarctic lakes have been studied under an Indian Antarctic expedition in a seasonal field campaign from December 2016 to February 2017. The lake water samples were collected from three surface freshwater lakes in the Larsemann Hills in East Antarctica named Nella, Progress and Reid. Every possible measurement has been taken during the sample collection and processing to avoid contamination. The water samples have been analysed for dissolved trace metal concentrations using Quadrupole-ICP-MS, macro nutrients (silicates, phosphates, nitrate and nitrite); analysis using auto analyser and chlorophyll-a (Chl-a) was measured using flourometer. Dissolved organic carbon (DOC) was analysed following standard methods [33]. Analytical precision within limits for each parameters and analysis was obtained.

2.4.2 Results

In this study, we have found that the major ion chemistry of the lakes in the Larsemann Hills is largely controlled by silicate weathering followed by atmospheric dust and seawater. Dissolved Cu plays an important role as a micronutrient that shows a strong correlation with DOC (**Figure 4**).

The sub-nanomolar concentration of trace metals indicates an insignificant impact of human activities on lake water chemistry, as also reported by earlier workers [34, 35]. Lakes are oligotrophic as indicated by Chl-a, and for most of the samples, the values are not more than 0.6 mg/m^3. The strong correlation between salinity, dissolved sulphates, dissolved Cu concentration, Chl-a and DOC plays the significant role of primary productivity, despite their lower concentration.

3. Conclusion

The chapter contributes to an understanding of the seasonal water balance of lakes located in Antarctic oases. The results and hydrological datasets from the technical reports were critically considered on the consistency of the variables measured as well as for uniforming methods and instruments applied during the measurement campaigns of 2011–2018. It was found that the hydrological observations remain more or less similar year by year for the methods and instruments used. However, the sets of the observed variables vary year by year even for a single lake; thus the multi-year datasets do not provide the similar dataset for the water balance components. The harmonisation of the seasonal hydrological programme is greatly needed to collect suitable datasets and to apply a modelling approach in climate-related and practical studies.

To prepare the uniformed programme of hydrological observations, the location of the temporal hydrological network was suggested for Lake Priyadarshini. The technical reports of the Finnish, Indian and Antarctic research programmes and previous publications allow for the recommendation of optimal solutions for the measurements, including locations of the gauges as well as a set of required instruments for the lakes Kitezh, Stepped, Nella, Progress and Reid.

The water balance is connected to the thermal and chemistry balances. The income and outcome terms of the water balance equation allow for evaluation of a lake retention time, which shows a strong correlation with trace metals like Zn, Cd, Co, Mn and Cu, thus playing an active role in lake water chemistry and primary productivity. The theoretical retention time is defined as a result of the division of a lake volume by inflow or outflow. In specific Antarctic conditions, the inflow/outflow on lakes only occurs during the summer. Thus, the retention time can be evaluated for warm and cold seasons separately and could be the topic of a future study.

Acknowledgements

We are supported by the Academy of Finland (contract number 304345) and Russian, Indian and Finnish Antarctic Research Program. We also acknowledge ESSO-National Centre for Antarctic and Ocean Research for the funding, logistic and analytical support and UGC for providing the research fellowship.

Conflict of interest

The authors declare that they have no conflicts of interest.

Author details

Elena Shevnina[1*], Ekaterina Kourzeneva[1] and Mohammad Nuruzzama[2]

1 Finnish Meteorological Institute, Helsinki, Finland

2 National Centre for Antarctic and Ocean Research, Goa, India

*Address all correspondence to: elena.shevnina@fmi.fi

References

[1] Bintanja R, van Oldenborgh GJ, Katsman CA. The effect of increased fresh water from Antarctic ice shelves on future trends in Antarctic Sea ice. Annals of Glaciology. 2015;**56**(69): 120-126. DOI: 10.3189/2015AoG69A001

[2] Shevnina E. FINNARP Field Operation Report [Internet]. 2018. Available from: http://www.antarctica.fi/documents/98503/0/FINNARP+vuosijulkaisu+2018.pdf/2ba9da09-1377-4737-901f-5aa7166b3ac9 [Accessed: July 14, 2018]

[3] Sokratova IN. Hydrological investigations in the Antarctic oases. Russian Meteorology and Hydrology. 2011;**36**:207. DOI: 10.3103/S1068373911030083

[4] Shevnina E, Kourzeneva E. Thermal regime and components of water balance of lakes in Antarctica at the Fildes peninsula and the Larsemann Hills. Tellus A: Dynamic Meteorology and Oceanography. 2017;**69**(1):1317202. DOI: 10.1080/16000870.2017

[5] Bromwich DH, Nicolas JP, Monaghan AJ. An assessment of precipitation changes over Antarctica and the Southern Ocean since 1989 in contemporary global reanalyses. Journal of Climate. 2011;**24**:4189-4209. DOI: 10.1175/2011JCLI4074.1

[6] Laybourn-Parry J, Wadham JL. Antarctic Lakes. Oxford: OUP; 2014. 40 p

[7] Chebotarev AI. Hydrology. Leningrad: Hydrometizdat; 1975. 544 p. (in Russian)

[8] Simonov IM. The Lakes of the Fildes Peninsula, King George (Vaterloo) Island. Inf. Bulletin Sovetskoy Antarkt. Expedicii. 1973;**85**:40-52. (in Russian)

[9] Simonov IM. Physical-geographic description of the Fildes peninsula (South Shetland Islands). Polar Geography. 1977;**1**(3):223-242. DOI: 10.1080/10889377709388627

[10] Delevendra S, Vimlesh P, Kamra A. Temperature-dependence of the positive intermediate ion concentrations at Maitri, Antarctica. Journal of Atmospheric and Solar–Terrestrial Physics. 2013;**104**:67-74. DOI: 10.1016/j.jastp.2013.08.011

[11] Singha J, Chatterjee I. Sixteenth Indian Expedition to Antarctica, Scientific Report, Department of Ocean Development, Technical Publication No. 14; 2000. 36 p

[12] Gillieson D, Burgess J, Spate A, Cochrane A. An Atlas of the Lakes of the Larsemann Hills, Princess Elizabeth Land, Antarctica. ANARE Research Notes No. 74. Kingston, Tasmania, Australia: The Publications Office, Australian Antarctic Division; 1990. 174 p

[13] Pickard J. Antarctic Oasis: Terrestrial Environments and History of the Vestfold Hills. Sydney: Academic Press; 1986. 368 p

[14] Seibert J. Conceptual runoff models—Fiction or representation of reality? [Comprehensive Summaries of Uppsala Dissertations from the Faculty of Science and Technology]. Uppsala University; 1999

[15] Grayson RB, Moore ID, McMahon TA. Physically based hydrologic modeling: 1. A terrain-based model for investigative purposes. Water Resources Research. 1992;**28**(10): 2639-2658. DOI: 10.1029/92WR01258

[16] Odrova T. Hydrophysics of Land Water Bodies. Leningrad: Gidrometeizdat; 1978. pp. 1-312. (in Russian)

[17] Girdiuk GV, Kirillova TV. Methodology to calculate the terms of radiation balance over the ocean. Meteorologiya i Gidrologiya. 1974;**12**: 63-69. (in Russian)

[18] Shine KP. Parametrization of the shortwave flux over high albedo surfaces as a function of cloud thickness and surface albedo. Quarterly Journal of the Royal Meteorological Society. 1984;**110**:747-764. DOI: 10.1002/(ISSN)1477-870X

[19] Mironov D, Golosov S, Heise E, Kourzeneva E, Ritter B, Scheider N, et al. FLake—A lake model for environmental applications. In: Folkard A, Jones I, editors. Proceedings of the 9th Workshop on Physical Processes in Natural Waters. Lancaster: Lancaster University; 2005

[20] Sabbe K, Hodson D, Verlryrn E, Udtaton A, Wilmotte A, Vanhoutte K, et al. Salinity, depth and the structure and composition of microbial mats in continental Antarctic lakes. Freshwater Biology. 2004;**49**:296-319

[21] Salo YA, Potahin MS, Tolstikov AV. Calculation of mean lake depth in absence of bathimetric observations an example of the Karelian lakes. Izvestija Russkogo Geograficheskovo Obschestva. 2010;**142**(3):43-47. (in Russian)

[22] Jawak SD, Luis AJ. A spectral index ratio-based Antarctic land-cover mapping using hyperspatial 8-band WorldView-2 imagery. Polar Science. 2013;**7**(1):18-38. DOI: 10.1016/j.polar.2012.12.002

[23] Shevnina E, Lesnicy L. Technical Report of Hydrological Studies in the Fildes Peninsula, the Season 57 RAE. 2012. 28 p

[24] Shevnina E, Vershinin K. Technical Report of Hydrological Studies in the Larsemann Hills, the Season 58 RAE. 2013. 22 p

[25] Fedorova I, Krasnov A, Zubov A. Technical Report of Hydrological Studies in the Larsemann Hills, the Season 57 RAE. 2012. 33 p

[26] Dvornikov Y, Evdokimov A: Technical Report of Hydrological Studies in the Larsemann Hills, the Season 62 RAE. 2017. 51 p

[27] Burba G. Eddy Covariance Method for Scientific, Industrial, Agricultural, and Regulatory Applications: A Field Book on Measuring Ecosystem Gas Exchange and Areal Emission Rates. Lincoln, NE, USA: LI-COR Biosciences; 2013. 331 p

[28] Naumov M: Technical Report of Hydrological Studies in the Larsemann Hills, the Season 59 RAE. 2014. 28 p

[29] Stüwe K, Braun HM, Peer H. Geology and structure of the Larsemann Hills area, Prydz Bay, East Antarctica. Australian Journal of Earth Sciences. 1989;**36**(2):219-241. DOI: 10.1080/08120098908729483

[30] Goudie AS, Viles HA. Weathering and the global carbon cycle: Geomorphological perspective. Earth-Science Reviews. 2012;**113**(1-2):59-71. DOI: 10.1016/j.erascirev.2012.03.005

[31] Walker JCG, Hays PB, Kasting JF. A negative feedback mechanism for the long term stabilization of Earth surface temperature. Journal of Geophysical Research. 1981;**86**(C10):9776. DOI: 10.1029/JC086iC10p09776

[32] West A, Galy A, Bickle M. Tectonic and climatic controls on silicate weathering. Earth and Planetary Science Letters. 2005;**235**(1-2):211-228. DOI: 10.1016/j.epsl.2005.03.020

[33] Clesceri LS, Eaton GA. Standard Methods for the Examination of Water and Wastewater. Washington, DC: American Public Health Association; 2003

[34] Burgess JS, Spate AP, Norman FI. Environmental impact of station development in the Larsemann Hills, Princess Elizabeth land, Antarctica. Journal of Environmental Management. 1992;**36**:287-299

[35] Gasparaon JSB. Human impact at Antarctica: Trace-element geochemistry of freshwater lakes in Larsemann Hills, East Antarctica. Environmental Geology. 2000;**39**(9):963-976

Chapter 6

The Climate of the Antarctic Peninsula during the Twentieth Century: Evidence from Ice Cores

Elizabeth R. Thomas and Dieter R. Tetzner

Abstract

The Antarctic Peninsula (AP) is a region of special climatological interest. The late twentieth century has been a period of warming surface temperatures, enhanced mass loss from melting glaciers and increased snowfall, which have a direct and measurable impact on global sea levels. However, the observational period for Antarctica is short. Observational records only began in the 1940s and much of our understanding of the wider spatial climate variability and glacial dynamics is limited to the satellite era (post 1979). Proxy records, such as those from ice cores, provide an invaluable tool to place these recent changes in context of the past few hundred years, allowing us to investigate climate variability over the entire twentieth century and beyond. In this chapter we review the climate of the AP during the twentieth century, as captured by the instrumental records, and extend our understanding of climate variability over the twentieth century based on climate proxies contained in ice cores. For this study we focus on stable water isotopes and snow accumulation and how they are influenced by changes in atmospheric circulation and sea ice conditions.

Keywords: Antarctic Peninsula, ice cores, climate, sea ice, surface mass balance

1. Introduction

The Antarctic Peninsula (AP) is the only landmass that cuts across the sub-Antarctic zone, transecting the circumpolar trough and forming a partial bridge between the Antarctic ice sheet and South America. Moreover, the AP forms a barrier to the strong south-westerly winds in the lower troposphere, separating the maritime climate from the Bellingshausen Sea from the continental climate on the Weddell Sea side [1–3]. These features highlight the AP as a unique observational platform to study the climatic interaction between low and high latitudes in the Southern Hemisphere. Indeed, climate records from this region add valuable information to understand how heat is transferred and how this has varied over time.

Instrumental observations in the AP are sparse and relatively short, only supported by a network of meteorological stations since the late 1940s [4]. The lack of long-term observational records hinders the possibility to put modern observations into a climatic perspective. However, ice cores retrieved from this region offer the possibility to evaluate climate over longer timescales. Chemical constituents and physical properties preserved in ice cores can be calibrated with meteorological and environmental observations to develop climatic reconstructions based on proxy records [5].

In this chapter, we evaluate the climate of the AP in recent decades, as expressed in the instrumental records, and highlight the important contribution that ice cores have made in expanding our understanding of climate variability over decadal to centennial timescales. We focus on two well-established ice core proxies (1) stable water isotopes, a proxy for past surface temperatures and (2) snow accumulation, a proxy for precipitation. We also demonstrate how both of these parameters have allowed us to investigate changes in large-scale atmospheric circulation beyond the instrumental period.

2. Climate of the Antarctic Peninsula during the twentieth century

2.1 Antarctic Peninsula ice cores

The AP is a region of high snow accumulation, capturing the seasonal deposition of chemical species in the ice [15], which when coupled with known volcanic horizons, allow high accuracy in the dating [16]. A significant advantage of the AP is that surface temperatures coincide with air temperature during cloud condensation, allowing a direct reconstruction of surface temperatures using stable water isotopes [17]. The best sites for ice core drilling are located on ice rises on the east coast, on ice domes, or on the central AP ice divide, where ice disturbance by deep flow is minimum [15, 17].

Since the mid-1970s, several ice cores have been drilled in the AP [6, 7, 9–11, 17–19]. The high annual accumulation, especially on the western coast, limits the temporal range of ice cores in this region, with only a few extending beyond the first half of the twentieth century (**Figure 1**; **Table 1**). The significantly different climatic regimes that prevail on each side of the AP makes it necessary to group these ice cores as west-coast regime and east-coast regime. In this chapter, we will focus our study on the ice cores retrieved from the western side of the AP.

2.2 Surface temperature

The surface temperature records are available from instrumental records at manned research stations and automatic weather stations, as global reanalysis data and obtained via remote sensing during the satellite era. In ice cores the stable water isotope record has been used as a proxy for past surface temperatures over centennial to millennial timescales.

2.2.1 Observations

Long-term trends in surface temperatures show that most of Antarctica has been warming since the records began [4]. The largest warming trends in the continent are concentrated on the western and northern parts of the AP [20–22], a region that exhibits the largest inter-annual variability in the whole continent [4]. Temperature measurements in Antarctica began in the early twentieth century [4] with the greatest density of records in the AP. The AP network includes the longest continuous Antarctic temperature record (Orcadas Station, South Orkney Islands), which extends back to 1903 [16]. After the International Geophysical Year (1957–1958), over a dozen permanent stations began to continuously record meteorological parameters. In addition to the station network, since the early 1980s, several Automatic Weather Stations (AWS) have been deployed in the region. These stations have helped to improve the spatial coverage of meteorological observations, as well as providing data from remote locations [4].

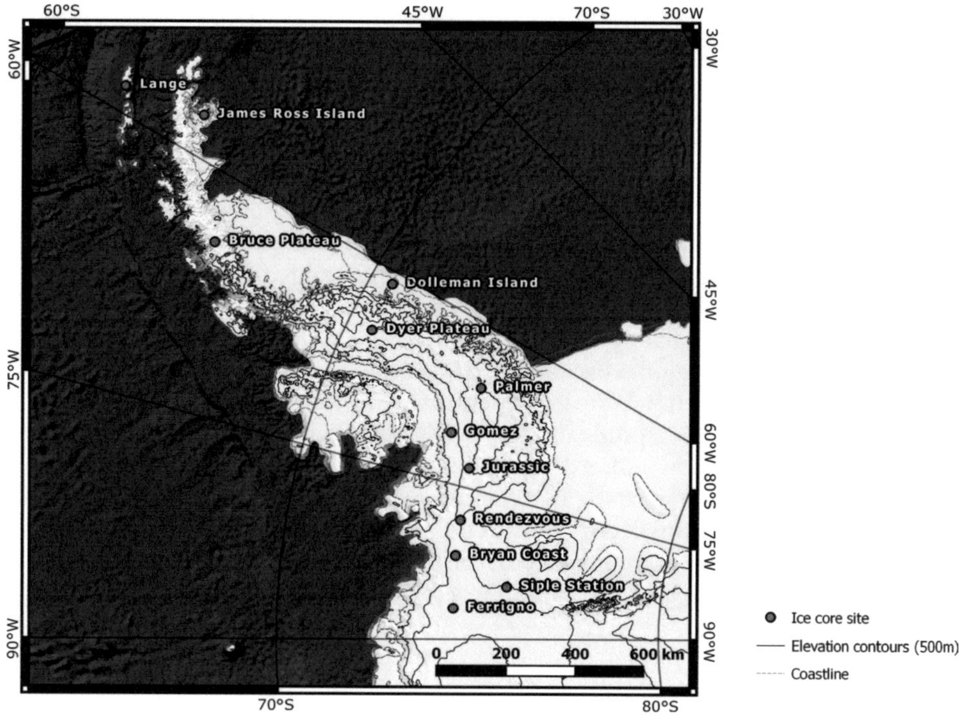

Figure 1.
Map of the Antarctic Peninsula showing the location of ice cores that extend back to the early twentieth century and beyond.

Site name	Longitude	Latitude	Elevation (m a.s.l)	Depth (m)	Data span (AD)	Data source
West coast regime						
Lange	−58.61	−62.11	690	50	1918–1995	[6]
Bruce Plateau	−64.07	−66.03	1976	448	1750–2009	[7, 8]
Dyer Plateau	−64.87	−70.67	2002	190	1504–1990	[9]
Palmer	−65.46	−73.86	1897	–	–	[10]
Gomez	−70.36	−73.59	1400	136	1858–2006	[11]
Jurassic	−73.06	−74.33	1139	–	–	[10]
Rendezvous	−78.16	−74.45	1006	–	–	[10]
Bryan Coast	−81.67	−74.49	1171	140	1712–2010	[10, 12]
Siple Station	−84.25	−75.92	1054	302	1410–1985	[13]
Ferrigno	−86.90	−74.57	1350	136	1703–2010	[10]
East coast regime						
James Ross Island	−57.68	−64.20	1542	364	~14,000 (year BP)–2007	[14]
Dolleman Island	−60.93	−70.58	398	133	1795–1986	[15]

Table 1.
Summary information for ice core sites.

The longest surface temperature record, from Orcadas Station, exhibits a trend of +0.21°C per decade since 1904 [22] with evidence from several stations of a significant warming since the early 1950s [22]. Most notably, the largest statistically significant trend of +0.54°C per decade observed on Faraday/Vernadsky station (1951–2011). Radiosonde data indicates that the largest warming has been confined to the lowest layers of the atmosphere (mainly the lower troposphere) [23]. The largest warming occurs during winter, with trends reaching up to +1.01°C per decade between 1950 and 2011 [24] and the greatest monthly temperature rise has been recorded in July (+1.7°C per decade between 1979 and 2007) [24]. Ultimately, the winter warming has caused an overall decrease in the annual temperature range and a change in the seasonal cycle [4].

Despite the strong regional warming trends measured in western AP in the late twentieth century, the annual mean temperature since the 1990s (1999–2014) has decreased at a statistically significant rate (<5% level), with the most rapid cooling during the summer season [25]. Additionally, Turner et al. [25] suggests that the rapid warming in the AP since the 1950s and subsequent cooling since the late 1990s are part of the large natural decadal-scale climate variability of the region. These findings highlight the need for longer surface temperature records to set the recent changes in a longer-term perspective and to assess the regional climate variability.

2.2.2 Ice cores

Ice cores from western AP provide climate records that extend back up to 600 years (**Table 1**). The linear relationship between local surface temperatures and stable isotopes in precipitation, at middle and high latitudes, allows us to reconstruct past surface temperatures from ice cores [26]. This approach has been applied at several locations in Antarctica over centennial to millennial timescales [27–30].

Isotopic temperature proxy data spanning the twentieth century is available from five ice core sites in the AP region (Bruce Plateau, Dyer Plateau, Gomez, Siple Station and Ferrigno) (**Figure 2**). A strong correlation between surface temperature measurements and temperatures reconstructed from ice cores (except for Dyer Plateau [9]), confirm the use of these records as valid proxies for local and regional temperatures on the west AP. A combination of these five ice cores provides a north–south transect along the AP which can be used to study latitudinal changes in the AP.

The five temperature proxy records do not provide a perfectly consistent picture of climate variability in the AP in the last five centuries. There are periods when most of the cores exhibit a similar temperature trend, but also periods when they show opposing trends. This probably reflects the degree to which ice core sites capture local and regional variability and the imprint of local high-frequency processes that complicate the interpretation [9]. The consistent feature is the recent rapid change in the isotopic composition, likely associated with the twentieth century warming trend measured in the meteorological record. For example, the reconstructed warming from Gomez ice core (+0.055°C per year) is consistent with the warming observed at Faraday/Vernadsky station (+0.054°C per year) (1955–2005) [31].

Overall, temperatures reconstructed from ice cores in the western AP show large inter-annual to inter-decadal variability [22, 31]. In particular, the onset of the warming since the 1950s in the southern cores (Gomez, Ferrigno and Siple Station) is delayed compared to the northern sites (Bruce Plateau and Dyer Plateau) and less pronounced at Siple Station [9, 13]. This could indicate that changes in the conditions in the northern AP did not impact the southern AP until some decades after [8]. Another consistent aspect among these records is a general cooling trend from ~1840/1850 to ~1920/1930 (absent in Siple Station ice core).

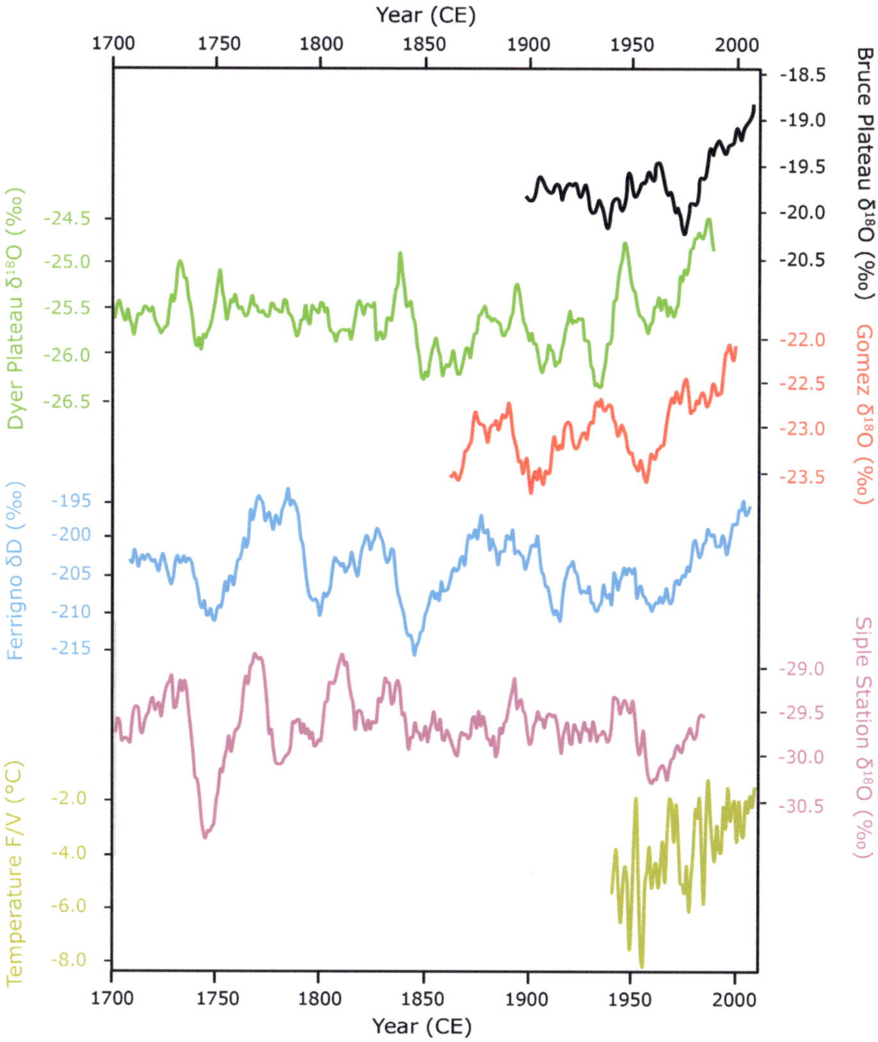

Figure 2.
11-year running mean isotope temperature proxies from ice cores in the AP region and annual mean surface temperature record from Faraday/Vernadsky Station (F/V), between 1700 and 2010.

Even though some records identify the last decades of the record as the warmest of the last centuries (Dyer Plateau and Gomez) [9, 31], others show larger warming trends and warmer decades occurring in the last centuries (Bruce Plateau, Ferrigno, Siple Station) [8, 9, 19]. In particular, in the Ferrigno ice core, Thomas et al. [19] reported larger 50-year warming trends occurring in the middle to late eighteenth century and in the middle nineteenth century. The analysis of the Ferrigno core revealed a reduction in the multi-decadal variability of surface temperatures during the twentieth century and suggested that the warming since the 1950s has not yet taken the system outside the natural range of climate variability [19].

Ice core temperature proxy records from the AP have provided evidence that the warming measured in the instrumental period is not just a local coastal phenomenon, but part of a regional warming trend covering the whole AP and extending back to the early twentieth century. Finally, they have proved that the current warming trends are not unprecedented in the last three centuries, suggesting that in some places the warming still remains within the range of natural range of climate variability.

Several authors have studied the possible drivers of the recent isotopic warming (see [30] and references therein) including the influence of atmospheric circulation and local sea ice conditions (see Section 2.4).

2.3 Snow accumulation

Antarctic surface mass balance (SMB) is the sum of all mass gains (snowfall) and mass losses (drifting snow erosion/deposition, sublimation and melt) from the surface of the Antarctic ice sheet.

2.3.1 Observations

Measuring SMB, in the AP is complicated as high precipitation and complex orography limit the availability of observational data. Remote sensing techniques such as Gravity Recovery and Climate Experiment (GRACE) [32] and radar back-scattering struggle to capture the small-scale features of the AP, while inverse methods of calculating SMB, including estimating mass discharge and elevation changes from satellite altimetry [33] require surface observations to correct for firn process.

Regional atmospheric climate models have proved reliable in simulating SMB over the Antarctic ice sheet [34, 35] and over the AP at high (14 km) resolution [36]. In addition, global reanalysis products have been used to approximate Antarctic SMB based on the spatial and temporal variability in precipitation-evaporation (P-E). The European Centre for Medium-range Weather Forecasts (ECMWF) ERA-40 reanalysis, has been shown to correlate with ice core accumulation records from West Antarctica [37], the southern Antarctic Peninsula [38] and across the majority of the Antarctic Peninsula [39]. The updated ECMWF reanalysis product, ERA-interim, has improved model physics with observational data supplemented by ECMWF's operational archives and been shown to represent snow accumulation at several sites across the southern AP and Ellsworth Land [10]. A recent study testing the performance of reanalysis and regional atmospheric climate model products, against over 3265 multiyear in situ observations, concluded that ERA-interim was the most reliable record of interannual precipitation compared with observations across the whole ice sheet [40]. While in terms of absolute snow accumulation observations the Regional Atmospheric Climate Model RACMO2.3 performed best.

RACMO2.3 is forced at its lateral atmospheric boundaries by ERA-interim reanalysis and at its lower ocean boundaries by sea ice fraction and sea surface temperature and been used to reconstruct AP SMB at ~5.5 km resolution for the period 1979–2014 [41]. The model reveals a large accumulation gradient across the AP, with precipitation on the western AP in excess of 3000 mm we year^{-1}, while the eastern AP receives less than 500 mm we year^{-1}. The average ice sheet integrated SMB, including ice shelves is estimated at 351 Gt year^{-1} [41].

During the observational period (1979–2014), there is no significant trend in either the reanalysis P-E data [40] or the modeled SMB from RACMO2.3 [41]. However, a recent compilation of Antarctic ice core snow accumulation records, regressed onto the SMB fields from RACMO2.3, concluded that SMB in this region has been increasing during the twentieth century [42, 43].

2.3.2 Ice cores

Antarctic SMB studies had largely dismissed the influence of the AP due to the lack of observational data. However, recent drilling efforts have greatly improved the spatial coverage in this region and demonstrated its importance in terms of total

Antarctic SMB. A compilation of all available ice core snow accumulation records revealed that SMB in the AP has increased at a rate of 12 Gt $year^{-1}$ since 1900 [42]. This equates to a 138 ± 58 Gt $year^{-1}$ (~20%) increase between the decadal average at the start of the twentieth century (1901–1910) and the decadal average at the start of the twenty-first century (2001–2010).

The dominant moisture source for the AP is the Amundsen Sea [10, 44], a region of high synoptic activity and the largest contributor to the total Antarctic meridional moisture flux [45]. High snow accumulation on the western AP is associated with reduced sea level pressure in the Amundsen Sea, leading to strengthened circumpolar westerlies and enhanced onshore flow of moist air masses originating from the mid-latitudes. Snow accumulation in the AP is related to the frequency of cyclones originating from low-latitudes over the South Pacific Ocean [46].

All AP ice cores reveal an increasing trend in snow accumulation during the twentieth century (**Figure 3**). The largest of which was reported at Gomez, where snow accumulation rates have doubled in the period 1850–2007 [11]. This positive trend is also evident as far south as the Ellsworth Land coast (Ferrigno and Bryan coast), but is not observed in other parts of west Antarctica which reveal little or no trend during this period [12]. The three southern ice core records are highly correlated with each other, and with reanalysis data and modeled SMB [42], suggesting they are representative of regional precipitation and SMB. Using a composite of the three records revealed that after 1919 the running decadal mean exceeds the baseline average and remains there for the entire twentieth century. The trend accelerates after 1984, when annual average snow accumulation values more than double that observed for the previous ~270 years [12]. The Bruce plateau ice core confirmed that the northern AP has also experienced an increase in snow accumulation during the late twentieth century, increasing at a rate of 0.19 mm we $year^{-1}$ since the 1950s [7], but the onset of the increase appears considerably later than that observed at the southern sites.

Prior to 1900 AD, the southern records (Bryan Coast, Ferrigno, Siple Station) suggest a period of relatively stable SMB, with a slight but not statistically significant negative trend (1750–1900). However, contrasting positive and negative trends (1750–1900) are observed at the two northern sites of Dyer Plateau and Bruce Plateau respectively [8].

2.4 The role of atmospheric circulation on AP climate

Atmospheric circulation describes the large-scale movement of air masses around the globe. It creates the winds and, together with oceanic circulation, is responsible for the transfer of the earth's thermal energy. Atmospheric circulation has a number of preferred modes of variability, such as El Nino Southern Oscillation (ENSO), the Southern Annular Mode (SAM) and the Pacific Decadal Oscillation (PDO), all of which influence the climate of the AP.

Major large-scale modes of atmospheric circulation have associated indices, calculated from meteorological data obtained from stations, climate re-analysis and from proxy records obtained from ice cores.

2.4.1 Observed changes in atmospheric circulation

The primary mode of atmospheric circulation in the Southern Hemisphere high latitudes is the SAM. This is a circumpolar pattern of atmospheric mass displacement which describes how the strength and location of the mid-to-high meridional pressure gradient change through time. These changes occur in a non-periodic way,

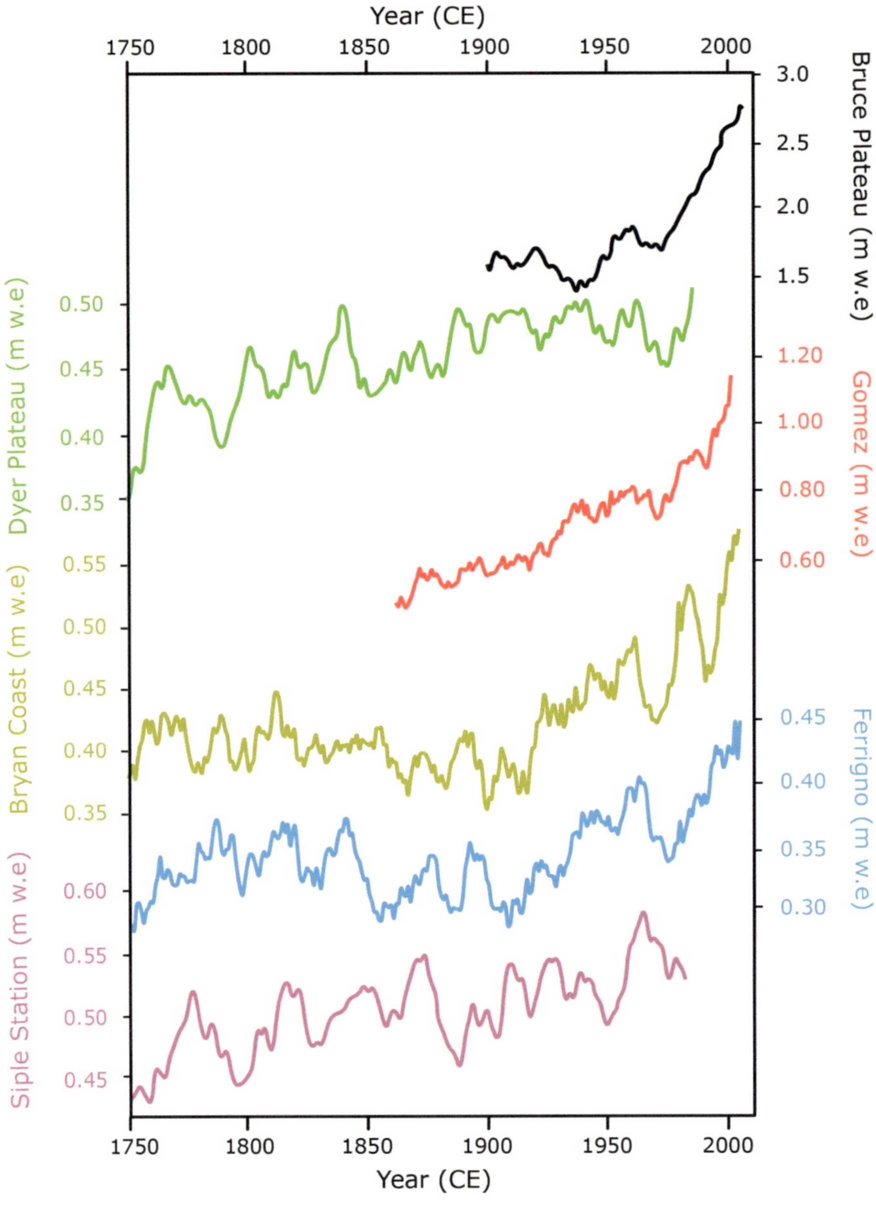

Figure 3.
11-year running mean of annual snow accumulation in ice cores from the AP region between 1750 and 2010.

varying within a range of days up to years [22]. Quantitatively, the SAM index is the difference of the zonal mean sea level pressure between records from six meteorological stations located in the mid-latitudes (around 40°S) and six stations in the Antarctic coast (around 65°S) [47]. A positive phase of SAM occurs when pressures around Antarctica are lower than pressures in the mid-latitudes. Conversely, a negative phase occurs when pressures over Antarctica are higher than pressures in the mid-latitudes.

Since the late 1970s, the SAM has trended into a positive phase, especially in the austral summer and autumn [47–49]. This has led to stronger circumpolar westerly winds over the Southern Ocean. In particular, during this period, the strength of the westerly winds has increased by 15–20% [50, 51] and it has been coupled with a poleward migration of the westerlies by 1–2° of latitude [22]. These two variations have impacted the cyclonic events south of 40°S, decreasing their frequency and increasing their intensity [52]. Furthermore, these new conditions have led to a deepening trend of the Amundsen Sea Low (ASL), a migrating climatological low-pressure center located over the Amundsen-Bellingshausen Seas, which strongly influences the climate in the Southern AP and the coast of West Antarctica [24]. The recent deepening trend of the ASL has produced the increase of meridional (onshore) winds that transport warm and moist air to the coast of Southern AP and West Antarctica, keeping this region mild compared to others at similar latitudes [53]. Overall, these changes caused by SAM, constitute one of the strongest climatic trends in the Southern Hemisphere over the last decades [54–56].

Another source of atmospheric circulation variability in the AP is ENSO, an inter-annual climatic variation over the tropical eastern Pacific Ocean which has impacts atmospheric conditions across the Pacific Basin and beyond [4]. The strength and phase of ENSO are commonly measured using the Southern Oscillation Index (SOI), calculated as the difference in sea level pressure between Tahiti Station and Darwin Station (Australia). Generally, ENSO events peak during September through February [57]. The way in which ENSO impacts the climate from the AP is through a teleconnection that causes a high latitude response, in the South Pacific-Drake Passage region, to tropical changes [57], intensifying climate variations depending on the SOI phase. In the last decades, ENSO teleconnection has shown marked decadal variability, presenting a weak teleconnection during the 1980s, while a strong teleconnection during 1990s [58, 59].

2.4.2 Influence of atmospheric circulation on stable water isotopes and snow accumulation

The recent strengthening in the circumpolar westerlies, associated with a positive phase of the SAM [48], has enhanced the meridional winds, drawing warm moist air to the western AP and influencing inter-annual temperature variability in ice core sites [19]. Indeed, at Gomez, approximately a third of the variability in annual mean surface temperatures (1957–2005) may be attributed to changes in the SAM [31] while it is responsible for a quarter of the snow accumulation increase (1957–2005). However, at all sites the relationship between SAM and both stable water isotopes and snow accumulation is not temporally stable.

Tropical sea surface temperatures (SST) influence atmospheric circulation in the Amundsen Sea region through the generation of a large-scale atmospheric wave train [60, 61]. This relationship is observed during the observational period (post 1979) as strong positive correlations between SMB in the AP and SSTs in the western tropical pacific associated with ENSO [12]. The snow accumulation records from the southern AP (Gomez, Ferrigno, Bryan Coast) reveal a strong negative ENSO-like pattern since 1980, which is not stable when extending the record to the past. Likewise, running decadal correlations of snow accumulation and SOI exhibit a positive correlation since 1980, but periods of insignificant and even negative correlations when extending over the full SOI record (1882–2010).

However, at the southern sites (Gomez, Ferrigno, Bryan Coast) snow accumulation does exhibit a positive correlation between tropical SSTs and surface pressure in the sub-tropical pacific that is not related to ENSO [19, 60]. This pattern appears stable back until at least ~1850s with the trend in snow accumulation, at least for the southern AP cores, consistent with reconstructed SSTs [12].

At Ferrigno, the stable water isotope record is positively correlated with proxy SSTs from coral growing at Rarotonga, in the sub-tropical pacific. The positive correlation remains throughout the past 240 years, with synchronous warm and cold periods observed [19]. This, together with the relationships observed between snow accumulation, suggests that changes in the tropical Pacific, not directly related to ENSO, are also driving high-latitude circulation.

In the northern AP, the Bruce Plateau snow accumulation record is modulated by climate variability in the tropical and subtropical regions, impacting this location through changes in the strength and position of the circumpolar westerlies [7]. The interplay between the phases of SAM, SOI and PDO proposed to explain the multi-decadal behavior between snow accumulation and large-scale atmospheric oscillations during the twentieth century.

An example of this is the relationship between the snow accumulation record and SAM indices. The relationship is positive and statistically significant ($R > 0.5$, $p < 0.001$) from 1971 to 2009, but not temporally stable over the last century, showing a sharp transition from positive to negative between 1950 and 1973 [7]. They explain this longer-term instability by changes in the strength of the tropical Pacific influence over the region. In particular, their results show that there is a stronger tropical Pacific (SOI) influence, over the snow accumulation record, when SAM and PDO are negative, while SOI remains on a positive phase (La Niña-event). Their results support the idea of an ENSO teleconnection modulated by SAM, but also by the phase of the PDO.

It appears that the coupling between modes of variability modulates snow accumulation in the AP [58] and may explain the acceleration in snow accumulation since the 1990s when both ENSO and SAM modes are in-phase.

The direct impact of changes in atmospheric circulation on parameters such as surface temperature and snow accumulation allow ice cores to record these changes over longer time scales. Providing their air-mass source region, or the transport pathway that they follow, is located within the region affected by the circulation changes. Backward trajectories studies have helped to determine the source region and transport pathways of all air masses reaching ice core locations [10, 38]. Even though the trajectories present a seasonal migration, their spatial coverage suggests that ice core records from this region are sensitive to changes in the ASL region [10] and the larger hemispheric scale atmospheric circulation (such as SAM and ENSO), which govern it [53]. Some ice cores from this region are better recording these changes and providing time series to study the atmospheric variability through time [7].

2.5 The role of sea ice on AP climate

Sea ice plays a major role in modulating global and regional climate. It alters the albedo of the Earth's surface and forms a barrier to the relatively warm surface ocean and the atmosphere above it. Changes in sea ice conditions can impact the availability of surface level moisture and the isotopic composition of air masses passing over it.

2.5.1 Observed changes in sea ice

Sea ice conditions are measured remotely, generated from brightness temperature data and passive microwave data collected by satellites. Sea ice conditions are commonly presented as (1) sea ice area, the portion of a grid cell covered by ice, (2) sea ice concentration (SIC), calculated as the percentage of ice cover within a 25 km^2 data cell or (3) sea ice extent (SIE), calculated as the northernmost latitude where sea ice concentration is 15% or greater.

During the observational period (1970 onwards), the total Antarctic sea ice area, calculated as the total area covered by ice, has increased [24, 62]. At a regional scale, however, there are marked differences. In the Weddell Sea and the Ross Sea sectors sea ice has increased, while the Bellingshausen Sea, and adjacent to the AP, there has been a significant decrease in sea ice.

2.5.2 Influence of sea ice on stable water isotopes and snow accumulation

The reduction in sea ice in the Bellingshausen and Amundsen Sea has been linked to the increased surface warming and increased snow accumulation on the western AP [7, 12, 63]. It has also been suggested that variations in sea ice can directly alter the isotopic composition of continental snow, based on the interaction between sea ice and surface exchange [64, 65]. The isotopic signal associated with water evaporated from the sea ice zone is believed to be deposited locally and thus the influence on stable water isotopes is expected to be greatest at coastal locations [65]. Indeed, at the Ferrigno site the relationship between stable water isotopes and sea ice in the Amundsen Ross sea is comparable with the relationship between stable water isotopes and site temperature [19].

Reduced sea ice results in enhanced availability of surface level moisture and increased poleward atmospheric moisture transport [45]. This results in greater snow accumulation, particularly at coastal sites, and this mechanism has been used to explain the longitudinal differences in AP snow accumulation trends during the twentieth century [12], with the greatest changes observed at sites where the adjacent sea ice decline is largest [62].

At Bruce Plateau, strong negative correlations exist between the observed sea ice extent in the Bellingshausen Sea and both stable water isotopes ($r = -0.55$) and snow accumulation ($r = -0.67$) [63]. Over the satellite era, Bellingshausen sea ice extent and snow accumulation exhibit significant decreasing and increasing trends, respectively, with sea ice extent explaining ~25% of the variance in snow accumulation at this site. The combined SMB composite produced from the AP ice cores, reveals a pattern of negative correlations with sea ice in the Bellingshausen Sea and positive correlations in the Amundsen-Ross Sea [42].

The Bruce Plateau snow accumulation record has been suggested as a proxy for past sea ice extent in the Bellingshausen Sea [63]. Porter et al., conclude that the increasing trend in accumulation since the 1970s suggests that the current rate of sea ice loss is unrivaled in the twentieth century. This is supported by other ice core proxy records such as methane sulfonic acid (MSA) record, a commonly used proxy for sea ice extent across Antarctica. The MSA record from the AP ice cores revealed a significant decline in sea ice in the Bellingshausen Sea during the twentieth century [66]. Conversely the MSA record from Ferrigno reflects changes in the Amundsen-Ross Sea, an area that is positively correlated with AP SMB [42, 67] and one that has exhibited a significant positive trend during the twentieth century.

Both the Bruce Plateau sea ice proxy based on snow accumulation and the Ferrigno sea ice proxy based on MSA, confirm the dominant role of ASL [63, 67], and hence large-scale modes of atmospheric variability, in driving changes in sea ice and ultimately AP SMB.

3. Conclusions

The Antarctic Peninsula has experienced considerable climate change during the twentieth century. The short observational period has provided compelling evidence of warming surface temperatures, increased glacial melt and mass loss [68] and reduced sea ice in the neighboring Bellingshausen Sea [62]. However, the observational period is short. A small number of meteorological observations span the past 50 years but the records are sparse and often dominated by local conditions. Here we have demonstrated the important role that ice cores have played in placing these recently observed changes in context.

Ice core stable water isotope records have demonstrated that the reported warming from stations in the northern AP since the 1950s is not just a local phenomenon, but part of a statistically significant 100-year regional warming trend [7, 31]. However, the ice core records also provide evidence that larger, more abrupt warming and cooling trends have occurred in recent centuries [19].

Ice core snow accumulation records represent mass gains to the ice sheet, a vital component of the total Antarctic mass balance. The observed ice melt in the AP since the 1990s [63] represents a mass loss, while the ice core records provide evidence of significant mass gain during the twentieth century [7, 11, 19]. Ice cores have provided evidence that SMB for the whole of Antarctica has increased since 1800, with the largest contribution (~75%) from the AP, where SMB has increased by 123 ± 44 Gt $year^{-1}$ [42].

The increase in surface temperature and SMB has been linked to changes in sea ice and atmospheric circulation. The observational records demonstrate a shift to the positive phase of the SAM since the 1957s that has increased the strength of the Southern Hemisphere westerly winds, deepened sea level pressures in the Amundsen Sea (ASL) and reduced sea ice in the Bellingshausen Sea. These later changes have also been attributed to the increased strength of ENSO, particularly since the 1990, with evidence interplay between these two modes is responsible for the acceleration in surface temperature and SMB in the late twentieth century.

The ice core records capture the influence of large-scale modes of climate variability over centennial time scales. They reveal that changes in SMB are sensitive to changes in the strength and phase of SAM, but that the relationship with ENSO is not temporally stable. However, the observed tropical teleconnection between climate on the AP and surface pressure and sea surface temperatures in the tropical pacific that are not related to ENSO [60], is consistent on centennial time scales [12, 19].

The observational records suggest that the interplay between modes of variability can have a considerable impact on climate of the AP [58]. Indeed, since the 1990s both SAM and ENSO have been in their positive phase, allowing for an amplification of the tropical teleconnection. In the ice core records the late twentieth century is characterized by a period of increased inter-annual variability and exceptionally high values in SMB [42] and sea ice [63–65], both of which are modulated by the variability in ASL (driven by SAM and ENSO). The combination of climate parameters and atmospheric circulation captured by the ice cores from the AP suggest that this recent coupling of SAM and ENSO is unprecedented in the past 300 years [12].

Acknowledgements

This work was funded by the British Antarctic Survey, part of the Natural Environment Research Council (NERC) and UK Research and Innovation (UKRI). D. Tetzner is funded on a CONICYT-Chile Cambridge scholarship.

Author details

Elizabeth R. Thomas* and Dieter R. Tetzner
British Antarctic Survey, Cambridge, UK

*Address all correspondence to: lith@bas.ac.uk

References

[1] Schwerdtfeger W. Climate of the Antarctic, climates of the polar regions. In: Orvig S, editor. World Survey of Climatology. Vol. 14. New York: Elsevier; 1970. pp. 253-355

[2] Martin P, Peel D. The spatial distribution of 10 m temperatures in the Antarctic Peninsula. Journal of Glaciology. 1978;**20**(83):311-317

[3] Schwerdtfeger P. Weather and climate of the Antarctic. In: Number 15 in Series Developments in Atmospheric Science. Amsterdam, New York: Elsevier; 1984

[4] King J, Turner J. Antarctic meteorology and climatology. In: Cambridge Atmospheric and Space Sciences Series. Cambridge, UK: Cambridge University Press; 1997. p. 409

[5] Legrand M, Mayewski P. Glaciochemistry of polar ice cores: A review. Reviews of Geophysics. 1997;**35**(3):219-243

[6] Simões JC, Ferron FA, Bernardo RT, Aristarain AJ, Stiévenard M, Pourchet M, et al. Ice core study from the king George island, south shetlands, Antarctica. Pesquisa Antártica Brasileira. 2004;**4**:9-23

[7] Goodwin BP, Mosley-Thompson E, Wilson AB, Porter SE, Sierra-Hernandez MR. Accumulation variability in the Antarctic Peninsula: The role of large-scale atmospheric oscillations and their interactions. Journal of Climate. 2016;**29**(7):2579-2596

[8] Goodwin BP. Recent Environmental Changes on the Antarctic Peninsula as Recorded in an Ice Core from the Bruce Plateau. Ohio, USA: The Ohio State University; 2013

[9] Thompson LG, Peel D, Mosley-Thompson E, Mulvaney R, Dal J, Lin P, et al. Climate since AD 1510 on dyer plateau, Antarctic Peninsula: Evidence for recent climate change. Annals of Glaciology. 1994;**20**:420-426

[10] Thomas ER, Bracegirdle TJ. Precipitation pathways for five new ice core sites in Ellsworth Land, West Antarctica. Climate Dynamics. 2015;**44**(7-8):2067-2078

[11] Thomas ER, Marshall GJ, McConnell JR. A doubling in snow accumulation in the western Antarctic Peninsula since 1850. Geophysical Research Letters. 2008;**35**(1):L01706. DOI: 10.1029/2007GL032529

[12] Thomas ER, Hosking JS, Tuckwell RR, Warren R, Ludlow E. Twentieth century increase in snowfall in coastal West Antarctica. Geophysical Research Letters. 2015;**42**(21):9387-9393

[13] Mosley-Thompson E, Thompson LG, Grootes PM, Gundestrup N. Little ice age (neoglacial) paleoenvironmental conditions at siple station, Antarctica. Annals of Glaciology. 1990;**14**:199-204

[14] Mulvaney R, Abram NJ, Hindmarsh RC, Arrowsmith C, Fleet L, Triest J, et al. Recent Antarctic Peninsula warming relative to Holocene climate and ice-shelf history. Nature. 2012;**489**(7414):141

[15] Peel D. Ice Core Evidence from the Antarctic Peninsula Region. In: Climate since AD. 1500. Bradley RS, Jones PD. editors. London: Routledge. 1992. pp. 549-571

[16] Mosley-Thompson E, Thompson LG. Ice core paleoclimate histories from the Antarctic Peninsula: Where do we go from here? In: Antarctic Peninsula Climate Variability: Historical and Paleoenvironmental Perspectives. Washington DC: AGU; 2003. pp. 115-127

[17] Peel D, Clausen H. Oxygen-isotope and total beta-radioactivity

measurements on 10 m ice cores from the Antarctic Peninsula. Journal of Glaciology. 1982;**28**(98):43-55

[18] Mulvaney R, Wolff EW. Spatial variability of the major chemistry of the Antarctic ice sheet. Annals of Glaciology. 1994;**20**:440-447

[19] Thomas ER, Bracegirdle TJ, Turner J, Wolff EW. A 308 year record of climate variability in West Antarctica. Geophysical Research Letters. 2013;**40**(20):5492-5496

[20] King J, Turner J, Marshall G, Connolley W, Lachlan-Cope T. Antarctic Peninsula climate variability and its causes as revealed by analysis of instrumental records. Antarctic Peninsula Climate Variability: Historical and Paleoenvironmental Perspectives. 2003;**79**:17-30

[21] Jones P, Marsh R, Wigley T, Peel D. Decadal timescale links between Antarctic Peninsula ice-core oxygen-18, deuterium and temperature. The Holocene. 1993;**3**(1):14-26

[22] Turner J, Barrand NE, Bracegirdle TJ, Convey P, Hodgson DA, Jarvis M, et al. Antarctic climate change and the environment: An update. Polar Record. 2014;**50**(3):237-259

[23] Marshall GJ, Lagun V, Lachlan-Cope TA. Changes in Antarctic Peninsula tropospheric temperatures from 1956 to 1999: A synthesis of observations and reanalysis data. International Journal of Climatology. 2002;**22**(3):291-310

[24] Turner J, Maksym T, Phillips T, Marshall GJ, Meredith MP. The impact of changes in sea ice advance on the large winter warming on the western Antarctic Peninsula. International Journal of Climatology. 2013;**33**(4):852-861

[25] Turner J, Lu H, White I, King JC, Phillips T, Hosking JS, et al. Absence of 21st century warming on Antarctic Peninsula consistent with natural variability. Nature. 2016;**535**(7612):411

[26] Dansgaard W. Stable isotopes in precipitation. Tellus. 1964;**16**(4):436-468

[27] Augustin L, Barbante C, Barnes PR, Barnola JM, Bigler M, Castellano E, et al. Eight glacial cycles from an Antarctic ice core. Nature. 2004;**429**:623-628

[28] Jouzel J, Masson-Delmotte V, Cattani O, Dreyfus G, Falourd S, Hoffmann G, et al. Orbital and millennial Antarctic climate variability over the past 800,000 years. Science. 2007;**317**(5839):793-796

[29] Masson-Delmotte V, Hou S, Ekaykin A, Jouzel J, Aristarain A, Bernardo R, et al. A review of Antarctic surface snow isotopic composition: Observations, atmospheric circulation, and isotopic modeling. Journal of Climate. 2008;**21**(13):3359-3387

[30] Stenni B, Curran MA, Abram NJ, Orsi A, Goursaud S, Masson-Delmotte V, et al. Antarctic climate variability on regional and continental scales over the last 2000 years. Climate of the Past. 2017;**13**(11):1609-1634

[31] Thomas E, Dennis P, Bracegirdle TJ, Franzke C. Ice core evidence for significant 100-year regional warming on the Antarctic Peninsula. Geophysical Research Letters. 2009;**36**(20):L20704. DOI: 10.1029/2009GL040104

[32] Tapley BD, Bettadpur S, Watkins M, Reigber C. The gravity recovery and climate experiment: Mission overview and early results. Geophysical Research Letters. 2004;**31**(9):L09607. DOI: 10.1029/2004GL019920

[33] Rignot E, Mouginot J, Scheuchl B. Ice flow of the Antarctic ice sheet. Science. 2011;**333**(6048):1427-1430

[34] Bromwich DH, Guo Z, Bai L, Chen Q-S. Modeled Antarctic precipitation. Part I: Spatial and temporal variability. Journal of Climate. 2004;**17**(3):427-447

[35] Lenaerts J, Van den Broeke M, van de Berg WJ, van Meijgaard E, Kuipers MP. A new, high-resolution surface mass balance map of Antarctica (1979-2010) based on regional atmospheric climate modeling. Geophysical Research Letters. 2012;**39**(4)

[36] Van Lipzig N, King J, Lachlan-Cope T, Van den Broeke M. Precipitation, sublimation, and snow drift in the Antarctic Peninsula region from a regional atmospheric model. Journal of Geophysical Research: Atmospheres. 2004;**109**(D24):D24106. DOI: 10.1029/2004JD004701

[37] Genthon C, Kaspari S, Mayewski P. Interannual variability of the surface mass balance of West Antarctica from ITASE cores and ERA40 reanalyses, 1958-2000. Climate Dynamics. 2005;**24**(7-8):759-770

[38] Thomas E, Bracegirdle T. Improving ice core interpretation using in situ and reanalysis data. Journal of Geophysical Research: Atmospheres. 2009;**114**(D20):D20116. DOI: 10.1029/2009JD012263

[39] Miles GM, Marshall GJ, McConnell JR, Aristarain AJ. Recent accumulation variability and change on the Antarctic Peninsula from the ERA40 reanalysis. International Journal of Climatology. 2008;**28**(11):1409-1422

[40] Wang Y, Ding M, Van Wessem J, Schlosser E, Altnau S, van den Broeke MR, et al. A comparison of Antarctic ice sheet surface mass balance from atmospheric climate models and in situ observations. Journal of Climate. 2016;**29**(14):5317-5337

[41] Van Wessem J, Ligtenberg S, Reijmer C, Van De Berg W, Van Den Broeke M, Barrand N, et al. The modelled surface mass balance of the Antarctic Peninsula at 5.5 km horizontal resolution. The Cryosphere. 2016;**10**(1):271-285

[42] Thomas ER, Melchior Van Wessem J, Roberts J, Isaksson E, Schlosser E, Fudge TJ, et al. Regional Antarctic snow accumulation over the past 1000 years. Climate of the Past. 2017;**13**(11):1491-1513

[43] Wang Y, Thomas ER, Hou S, Huai B, Wu S, Sun W, et al. Snow accumulation variability over the West Antarctic ice sheet since 1900: A comparison of ice core records with ERA-20C reanalysis. Geophysical Research Letters. 2017;**44**(22):11,482-111,90

[44] Turner J, Harangozo SA, Marshall GJ, King JC, Colwell SR. Anomalous atmospheric circulation over the Weddell Sea, Antarctica during the austral summer of 2001/02 resulting in extreme sea ice conditions. Geophysical Research Letters. 2002;**29**(24):2160. DOI: 10.1029/2002GL015565

[45] Tsukernik M, Lynch AH. Atmospheric meridional moisture flux over the Southern Ocean: A story of the Amundsen Sea. Journal of Climate. 2013;**26**(20):8055-8064

[46] Hosking JS, Fogt R, Thomas ER, Moosavi V, Phillips T, Coggins J, et al. Accumulation in coastal West Antarctic ice core records and the role of cyclone activity. Geophysical Research Letters. 2017;**44**(17):9084-9092

[47] Marshall GJ. Trends in the southern annular mode from observations and reanalyses. Journal of Climate. 2003;**16**(24):4134-4143

[48] Marshall GJ, Orr A, Van Lipzig NP, King JC. The impact of a changing Southern Hemisphere annular mode on Antarctic Peninsula summer temperatures. Journal of Climate. 2006;**19**(20):5388-5404

[49] Thompson DW, Solomon S, Kushner PJ, England MH, Grise KM, Karoly DJ. Signatures of the Antarctic ozone hole in Southern Hemisphere surface climate change. Nature Geoscience. 2011;**4**(11):741

[50] Korhonen H, Carslaw KS, Forster PM, Mikkonen S, Gordon ND, Kokkola H. Aerosol climate feedback due to decadal increases in Southern Hemisphere wind speeds. Geophysical Research Letters. 2010;**37**(2):L02805. DOI: 10.1029/2009GL041320

[51] Turner J, Marshall GJ. Climate Change in the Polar Regions. Cambridge, UK: Cambridge University Press; 2011

[52] Simmonds I, Keay K, Lim E-P. Synoptic activity in the seas around Antarctica. Monthly Weather Review. 2003;**131**(2):272-288

[53] Hosking JS, Orr A, Marshall GJ, Turner J, Phillips T. The influence of the Amundsen–Bellingshausen seas low on the climate of West Antarctica and its representation in coupled climate model simulations. Journal of Climate. 2013;**26**(17):6633-6648

[54] Thompson DW, Solomon S. Interpretation of recent Southern Hemisphere climate change. Science. 2002;**296**(5569):895-899

[55] Gille ST. Decadal-scale temperature trends in the Southern Hemisphere Ocean. Journal of Climate. 2008;**21**(18):4749-4765

[56] Young I, Zieger S, Babanin AV. Global trends in wind speed and wave height. Science. 2011;**332**(6028):451-455

[57] Fogt RL, Bromwich DH. Decadal variability of the ENSO teleconnection to the high-latitude South Pacific governed by coupling with the southern annular mode. Journal of Climate. 2006;**19**(6):979-997

[58] Clem KR, Fogt RL. Varying roles of ENSO and SAM on the Antarctic Peninsula climate in austral spring. Journal of Geophysical Research: Atmospheres. 2013;**118**(20):11,481-111,92

[59] Fogt RL, Bromwich DH, Hines KM. Understanding the SAM influence on the South Pacific ENSO teleconnection. Climate Dynamics. 2011;**36**(7-8):1555-1576

[60] Ding Q, Steig EJ, Battisti DS, Küttel M. Winter warming in West Antarctica caused by central tropical Pacific warming. Nature Geoscience. 2011;**4**(6):398

[61] Lachlan-Cope T, Connolley W. Teleconnections between the tropical Pacific and the Amundsen-Bellinghausens Sea: Role of the El Niño/southern oscillation. Journal of Geophysical Research: Atmospheres. 2006;**111**(D23). DOI: 10.1029/2005JD006386

[62] Turner J, Comiso JC, Marshall GJ, Lachlan-Cope TA, Bracegirdle T, Maksym T, et al. Non-annular atmospheric circulation change induced by stratospheric ozone depletion and its role in the recent increase of Antarctic Sea ice extent. Geophysical Research Letters. 2009;**36**(8):L08502. DOI: 10.1029/2009GL037524

[63] Porter SE, Parkinson CL, Mosley-Thompson E. Bellingshausen Sea ice extent recorded in an Antarctic Peninsula ice core. Journal of Geophysical Research: Atmospheres. 2016;**121**(23):13,886-13,900

[64] Bromwich DH, Weaver CJ. Latitudinal displacement from main moisture source control $\delta^{18}O$ of snow in coastal Antarctica. Nature. 1983;**301**:145-147

[65] Noone D, Simmonds I. Sea ice control of water isotope transport

to Antarctica and implications for ice core interpretation. Journal of Geophysical Research. 2004;**109**:D07105. DOI: 10.1029/2003JD004228

[66] Thomas ER, Abram NJ. Ice core reconstruction of sea ice change in the Amundsen-Ross seas since 1702 AD. Geophysical Research Letters. 2016;**43**(10):5309-5317

[67] Abram NJ, Thomas ER, McConnell JR, Mulvaney R, Bracegirdle TJ, Sime LC, et al. Ice core evidence for a 20th century decline of sea ice in the Bellingshausen Sea, Antarctica. Journal of Geophysical Research: Atmospheres. 2010;**115**(D23):D23101. DOI: 10.1029/2010JD014644

[68] The IMBIE Team. Mass balance of the Antarctic ice sheet from 1992 to 2017. Nature. 2018;**558**:219-222

Chapter 7

Characteristic Infrasound Events Associated with Sea-Ice Discharges in the Lützow-Holm Bay of Antarctica: April 2016

Takahiko Murayama, Masaki Kanao and Masa-Yuki Yamamoto

Abstract

Infrasound waves detected in Antarctica contain information on the physical interaction among the surface environment at the margin of the continent and surrounding ocean. Time-space variation of source location for infrasound excitation during mid-April 2016 was investigated by using a combination of two arrays deployed along the coast of the Lützow-Holm Bay (LHB), East Antarctica. The infrasound array observations detected temporal variations in distance from the sources and propagation direction. A few tens of infrasound events were identified during 10 days of the period, and many of them located in the northward direction from the array stations were inside the LHB and offshore in the Southern Indian Ocean. Many of the events had predominant frequency content of few Hz, which were higher than microbaroms generated from the ocean. By comparing with MODIS satellite image at the same period, these sources were considered to be the ice-related phenomenon associated with the discharge of fast sea ice from the LHB.

Keywords: infrasound, array analysis, sea-ice discharge, Lützow-Holm Bay, Antarctica, cryosphere dynamics

1. Introduction

"Infrasound" is a sub-audible pressure wave with frequency content from the cutoff of a sound (3.21 mHz, for a 15°C atmosphere) to the lowest of the human audible band (20 Hz) and propagates thousands of kilometers along the Earth's surface by considerable excitation energy [1]. There are many examples of infrasound excitation by a couple of plausible sources: volcanic eruptions, oceanic swells, large earthquakes, aircrafts, thunder and sprites, fireballs, meteoroid, reentry of artificial vehicles, aurora activities, etc. [2–6].

In the polar region, time-space variations in atmospheric pressure are generated by physical interaction among multi-spheres (atmosphere, oceans, cryosphere, and the surface of the solid earth). These interactions are actively involved in the surface environment, and their unknown sources can be measured by infrasound networks deployed in the polar region (**Figure 1**). Recently, [7] conducted simultaneous observation by both seismic and infrasound sensors at the Bowdoin Glacier in

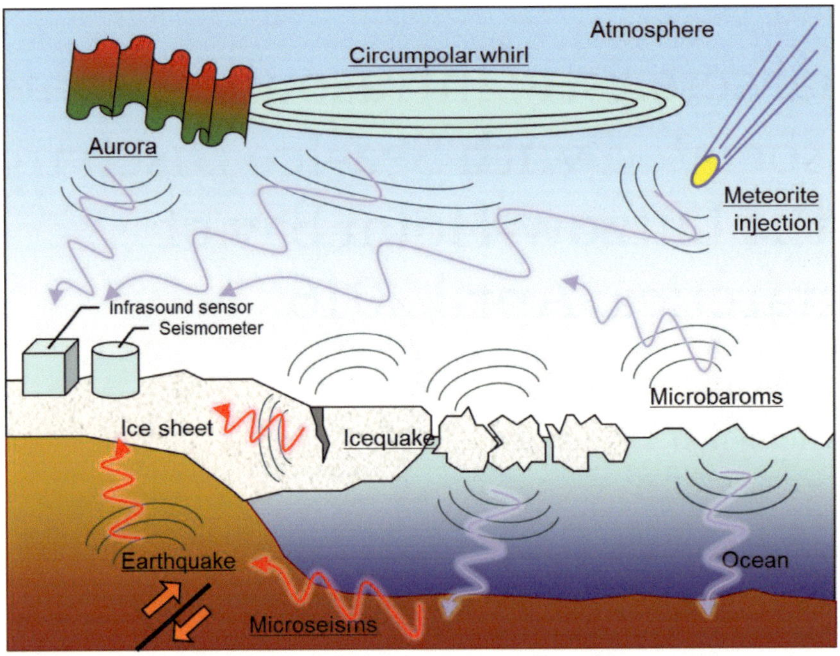

Figure 1.
Schematic illustration of interdisciplinary physical interaction within multi-spheres in the Antarctic. Several kinds of seismic, infrasonic, and hydroacoustic waves are generated by surface environmental sources (after [12]).

Greenland; they found ground validate infrasound sources very precisely by using time-lapse cameras and better-localized sources due to their small size.

In April 2008, infrasound observation was started by using a single sensor at the main Japanese station, Syowa (SYO; 69.0S, 39.6E), in the Lützow-Holm Bay (LHB) of Antarctica. The single infrasound sensor at SYO has been continuously recording the data over the seasons since 2008 and has clearly marked the background contamination signals of oceanic swells (microbaroms) [8], and the first 3-year variability of frequency contents and power spectrograms had been investigated [9].

In austral summer 2013–2014, several field stations of infrasound sensors were established along the coast of LHB (**Figure 2**). In particular, two infrasound arrays with different diameters were deployed both on the outcrop at SYO and the continental ice sheet (S16) near the eastern coast of LHB [10]. They reported the frequency contents and source orientation of the microbaroms from the Southern Ocean for 50 days of data in austral summer in 2013. During the initial observation period of the arrays, as one of the examples of unique data, airburst shock waves emanating from a meteor entering the atmosphere over the Russian Republic were identified on February 15, 2013.

Three infrasound events in 2015 were also identified by using the two arrays deploying at SYO and S16, and the excitation source locations were determined along the coast, inside the sea ice and surrounding the islands [11]. Moreover, the long-term variability of the source location of the infrasound excitation for 8 months in January–August 2015 was investigated by using the same arrays [12]. Plausible source mechanisms of these events were estimated regarding the surface environmental change, in particular, cryosphere dynamics surrounding the LHB. Especially focusing on the facts on mid-April, 2015 were studied associated with the discharge of sea ice together with the collision with icebergs surrounding the LHB.

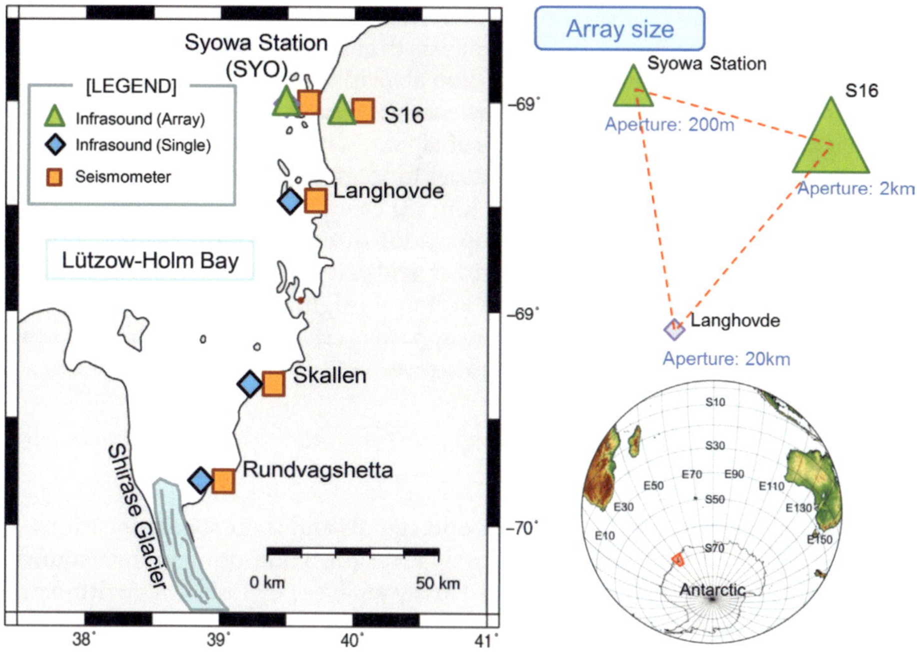

Figure 2.
(Left) locations of array deployment in the Lützow-Holm Bay (LHB). Array stations of infrasound (green triangles), single stations of infrasound (blue diamond), and broadband seismometers (orange squares) are shown. (Upper right) Array configuration of infrasound stations so as to localize the source signals. Tripartite arrays have been deployed by small size (at Syowa Station (SYO); aperture of 200 m), medium size (at S16, S17, P50; aperture of 2 km), and large size (combination of other outcrop stations such as Langhovde, aperture of 20 km or larger).

In this chapter, in addition to these previous works in the target region of LHB, time-space variation of source location for infrasound excitation during mid-April 2016 was investigated by using a combination of two arrays deployed along the coast of LHB, when a significant volume of sea ice was discharged from the Bay to the Southern Ocean.

2. Observation and analysis

A total of nine infrasound instruments have been carried out along the eastern coast of LHB since January 2013 [10]. Two triangle array alignments with different diameters were constructed at SYO (with a 100 m spacing triangle) and the vicinity of S16 over ice sheet (1 km space triangle) where 15 km eastward location from SYO (**Figure 2**). Besides the arrays, single sensor stations have been set at several outcrops in LHB (Langhovde, Skallen, and Rundvagshetta). These complicate that array configuration of these stations was adopted to localize detected signals efficiently by recognizing identical wavelengths with corresponding frequencies for each array size.

The Chaparral Physics microbarometer (Model 25, detectable frequency of 0.1–200 Hz) has been utilized in most of the stations. In these stations, hose arrays were aligned to reduce wind noises by mechanical low-pass filtering [13, 14]. Multiply-connected porous hoses are adopted at the SYO array; in contrast, a single array configuration was utilized in the other stations so as to help simplify their installations. The porous hoses are buried beneath the mounds of stones/snow ice collected from around the observation sites to reduce the vibration effect of winds. Detail configuration of the observation system is given in [10].

In order to speculate propagation directions and locations of infrasound events, multiple-step approach was used; the first was to generate a catalog for each array using a progressive multichannel correlation algorithm (PMCC) [15, 16]; the second was by suing two bulletins of the arrays (SYO and S16). A flowchart of array analysis is as follows: (1) search the pairs of signals within ±80 s of detected time difference on the basis of the bulletin dataset for both arrays on the basis of the distance of two arrays which is about 20 km; (2) calculate the cross point by using a spherical triangle method based on propagation direction and apparent velocities for each array; (3) set the candidate origin as grids around the cross point within ±5 degree range, followed by calculating averaged origin time, and select the most probable grid; and (4) evaluate calculated apparent velocities (V) within the range of (0.28 m/s $\leqq$ V $\leqq$ 0.36 m/s) for both the arrays.

3. Results and discussion

Several examples of detected infrasound signals and their source locations from array analyses are demonstrated in this section. Time-domain infrasound waveforms for three events and resultant array analyses by PMCC algorithm are shown in **Figure 3**. The identified infrasound signals were presumably associated with a series of discharge events of sea ice from LHB. Three examples of the detected waveforms are (1) April 3, 2016 (forecasted origin time, 03:27 UTC; 3 minutes of data duration), (2) April 11, 2016 (forecasted origin time, 22:00 UTC; 6 minutes of data duration), and (3) April 7, 2016 (forecasted origin time, 16:42 UTC; 6 minutes of data duration), respectively. The upper panels of each waveform represent the results of PMCC analysis: back azimuth (station-to-source) direction and the apparent velocities. The lower panels represent band pass-filtered infrasound waveforms observed at SYO array. Green broken squares on the waveforms correspond to the time windows of the detected infrasound events. All three events came from northwest direction to the SYO array as identified from the back azimuth distributions, which correspond to the inside LHB or offshore in South Indian Ocean.

Since these waveforms have relatively high-frequency contents of few Hz and amplitudes between 0.01 and 0.1 Pa, they are not considered to be oceanic swell origins propagating from Southern Ocean (microbaroms; dominant frequency around 0.2 Hz [9]). Ice quakes, calving of glaciers and ice cliffs, the collision of icebergs and sea ice, and other candidates of generating sources involving cryosphere dynamics could produce infrasound signals with sufficient energy that is recordable to the arrays. These kinds of cryoseismic sources are likely to reflect short-term variations in the surrounding environment, and their temporal change may provide indirect evidence of climate change at the local polar region [17]).

These detected infrasound events with high-frequency contents on mid-April 2016 were compared with the MODIS satellite images (provided by NASA) which give cryosphere information at the target area. **Figure 4** represents the MODIS images around LHB on April 3 (upper left), April 5 (upper right), April 8 (lower left), and April 9 (lower right), 2016, respectively. Discharged areas of fast sea ice from LHB are circled by broken red line. Distribution of fragmentations of sea ice (fast ices) and several tips of icebergs are identified outside the Bay. Moreover, other geophysical data of tide gauges deploying at SYO recorded the large amplitude of oceanic swells on April 10 associated with the discharge events during the mid-April period.

Array analysis results of estimated source locations of infrasound excitation overlapping on MODIS image (April 8) during the period from April 3 to April

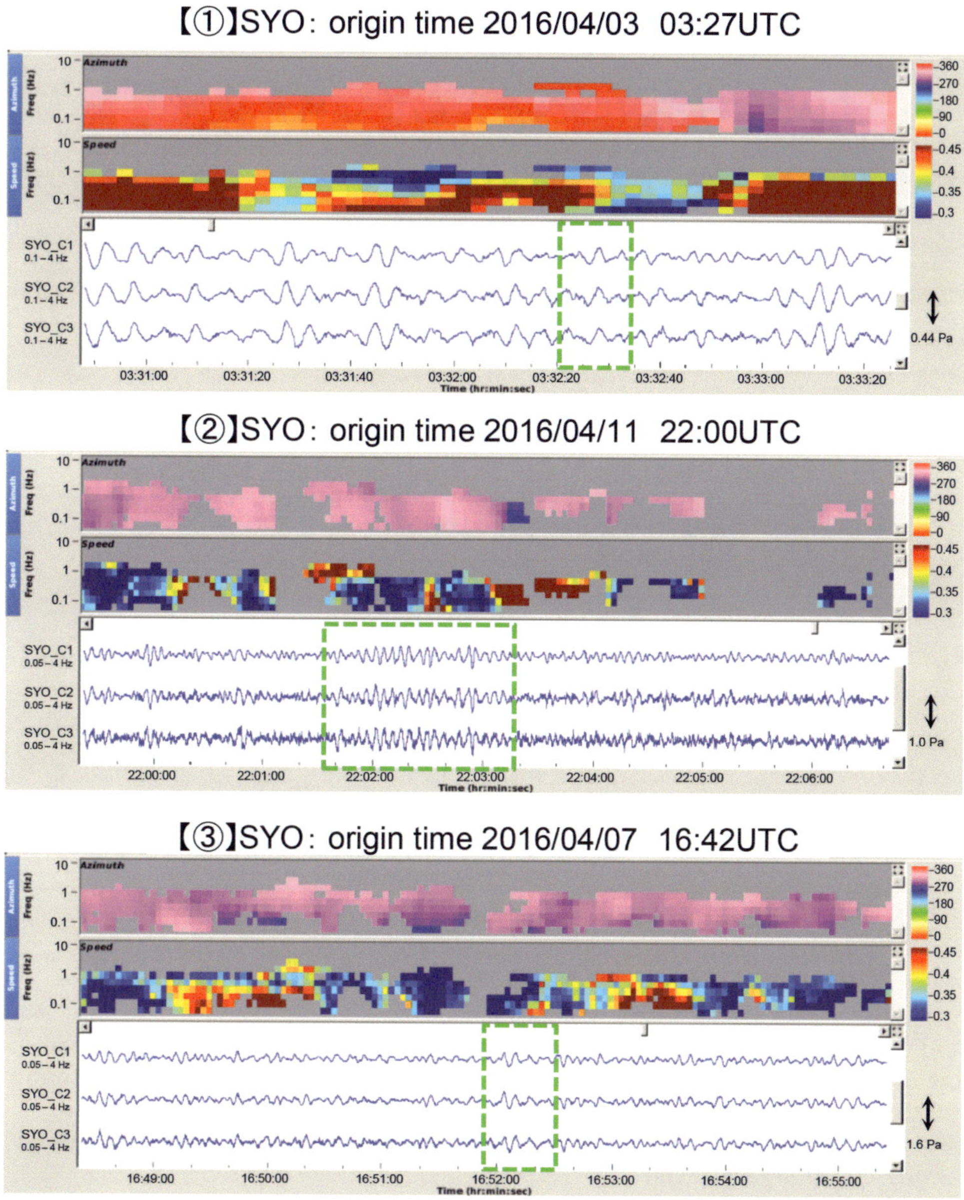

Figure 3.
Detection of infrasonic signals associated with discharge events of sea ices from LHB on (1) April 3, 2016 (forecasted origin time, 03:27 UTC), (2) April 11, 2016 (forecasted origin time, 22:00 UTC), and (3) April 7, 2016 (forecasted origin time, 16:42 UTC). The upper panels show the results of PMCC analysis: Back azimuth (station-to-source) direction and the apparent velocities. The lower panels represent band pass-filtered waveforms observed at SYO array. Green broken squares on the waveforms correspond to the time windows of detected infrasound events.

12 are presented in **Figure 5**. Estimated source locations (N = 175) are colored according to the time from the beginning of April 3. Three events correspond to the waveform examples shown in **Figure 4** which are raveled by red circled numerous. Majority of the infrasound sources were located in the northwest direction from LHB, which appeared to extend toward the Indian Ocean. Besides, more than 10 events are recognized to be located toward the north direction from LHB. Majority of the events are considered to concentrate on the vicinity and the edge of icebergs

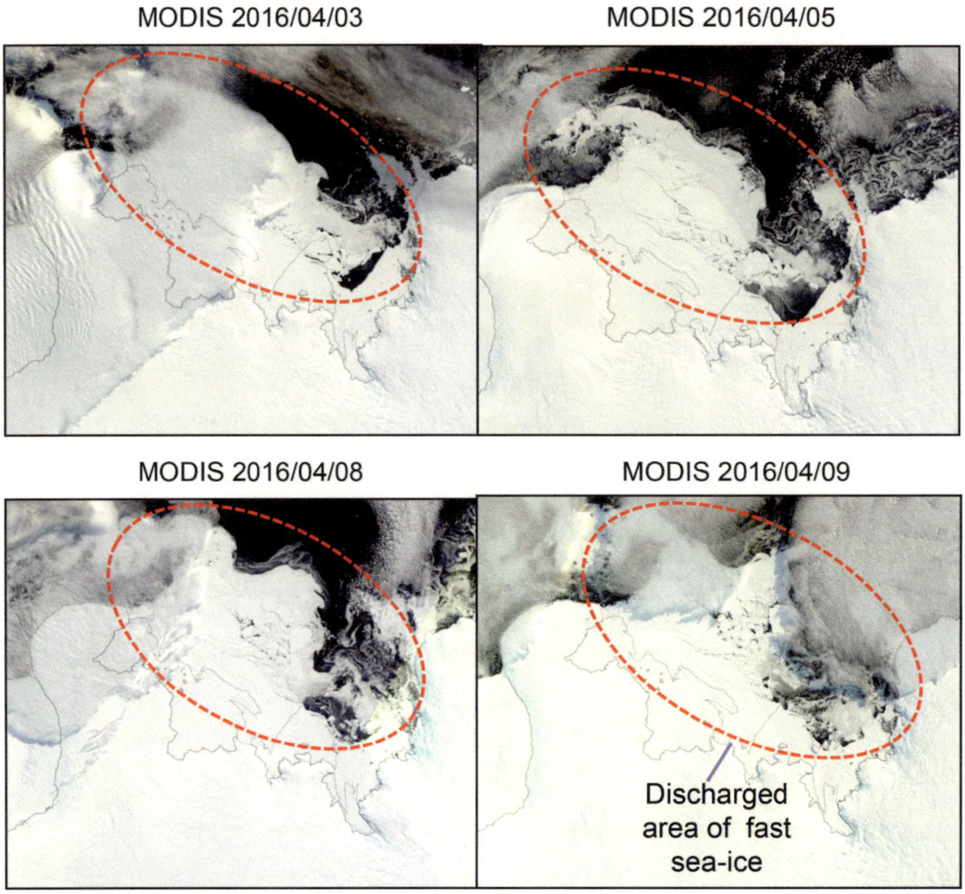

Figure 4.
MODIS satellite images around LHB on April 3 (upper left), April 5 (upper right), April 8 (lower left), and April 9 (lower right), 2016, respectively. Discharged areas of the fast sea ices are circled by red broken lines.

and the pack ice area where cryospheric dynamic movements (crashing, collision, break-off, etc.) involving oceanic swells were considered to generate the infrasound sources. Besides, many of the events occurred after April 11 when the last stage of sea-ice discharge events, just after the large oceanic swells, arrived at SYO as recorded by tide gauge data.

Regarding the effect of oceanic swells, microbaroms varied significantly both in amplitude and frequency contents in relatively long-period waveforms. These variations correspond to local atmospheric conditions, which are related to prevailing weather conditions in the vicinity of the area. As discussed in [12], infrasound signals below 3 Hz frequency content are supposed to contain in some extent the "microbaroms" which can be excited by storms during whole season particular in austral winter. By conducting this study, the array alignment of infrasound stations in LHB was efficiently operating, and it provides precise information on the arrival direction of infrasound excitation sources. The exact location of the generation sources might be compared with those obtained by seismic/geophysical approaches in the future. The oceanic-atmospheric coupling effects on infrasound could be explained by the relationship among complex Earth system in the polar environment. In this regard, infrasound monitoring in the Antarctic is a new proxy for detecting local/regional environmental change within global climate variation.

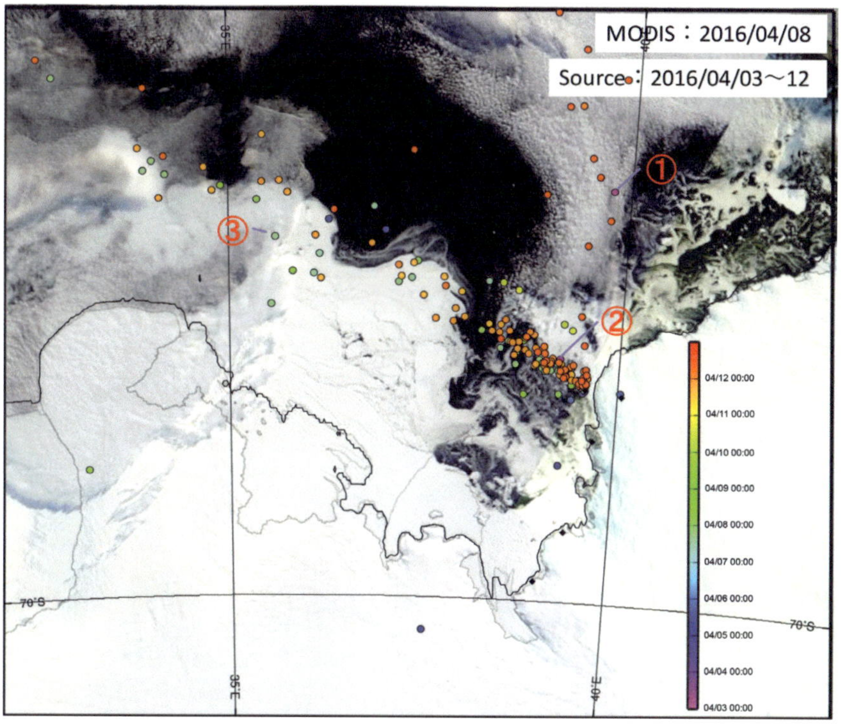

Figure 5.
*Result of estimated source locations of infrasound excitation with MODIS image (April 8) during the period from April 3 to April 12, 2016, by using two arrays (SYO and S16). Black solid diamonds are the infrasound stations in LHB. The estimated source locations (N = 175) are colored according to the time from the beginning of April 3. Three events correspond to the waveform examples (**Figure 4**) are which raveled by red circled numerous.*

4. Conclusion

Time-space variation of source location for infrasound excitation in mid-April 2016 was investigated by using two pair arrays deployed in LHB. A few tens of infrasound sources were determined during the period, and a majority of these events was located in northward orientation inside the sea-ice distributing area, attributed to the cryosphere origins such as the discharge of fast sea ice and collisions with icebergs by comparison with MODIS data. Many of the identified events had frequency contents of few Hz, which were higher than microbaroms from oceanic swells. Continuous observation of infrasound could be a proxy for monitoring surface environment and currently be going on climate change in the Antarctic.

Acknowledgements

The authors express their appreciation to many collaborators regarding infrasound observations around LHB, in particular, the members of the Japanese Antarctic Research Expeditions (JARE). The authors also acknowledge the National Aeronautics and Space Administration (NASA) for the utilization of MODIS images. This work was supported by JSPS KAKENHI Grant Number 26241010 (PI by Dr Masaki Kanao).

Author details

Takahiko Murayama[1*], Masaki Kanao[2] and Masa-Yuki Yamamoto[3]

1 Japan Weather Association, Tokyo, Japan

2 National Institute of Polar Research, Research Organization of Information and Systems, Tachikawa-Shi, Tokyo, Japan

3 Kochi University of Technology, Kami-Shi, Kochi, Japan

*Address all correspondence to: murayama@jwa.or.jp

References

[1] Hedlin M, Garces M, Bass H, Hayward C, Herrin G, Olson JV, et al. Listening to the secret sounds of Earth's atmosphere. Eos, Transactions American Geophysical Union. 2002;**83**(557):564-565

[2] Arai N, Iwakuni M, Watada S, Imanishi Y, Murayama T, Nogami M. Atmospheric boundary waves excited by the tsunami generation related to the 2011 great Tohoku-Oki earthquake. Geophysical Research Letters. 2011;**38**:L00G18. DOI: 10.1029/2011GL049146

[3] Matoza RS, Hedlin MAH, Garces MA. An infrasound array study of Mount St. Helens. Journal of Volcanology and Geothermal Research. 2007;**160**:249-262

[4] Arrowsmith SJ, ReVelle DO, Edwards WN, Brown P. Global infrasonic signals from three large bolides. Earth, Moon and Planets. 2008;**102**:357-363

[5] Le Pichon A, Blanc E, Drob D. Probing high-altitude winds using infrasound. Journal of Geophysical Research. 2005;**110**:D20104. DOI: 10.1029/2005JD006020

[6] Wilson CR. Auroral infrasonic waves. Journal of Geophysical Research. 1969;**747**:1812-1836. DOI: 10.1029/JA074i007p01812

[7] Podolskiy EA, Genco R, Sugiyama S, Walter F, Funk M, Minowa M, et al. Seismic and infrasound monitoring of Bowdoin glacier, Greenland. Low Temperature Science. 2017;**75**:15-36. DOI: 10.14943/lowtemsci.75.15

[8] Yamamoto M-Y, Ishihara Y, Kanao M. Infrasonic waves in Antarctica: A new proxy for monitoring polar environment. International Journal of Geosciences. 2013;**4**:797-802. DOI: 10.4236/ijg.2013.2800498

[9] Ishihara Y, Kanao M, Yamamoto M-Y, Toda S, Matsushima T, Murayama T. Infrasound observations at Syowa Station, East Antarctica: Implications for detecting the surface environmental variations in the polar regions. Geoscience Frontiers. 2015;**6**:285-296

[10] Murayama T, Kanao M, Yamamoto M-Y, Ishihara Y, Matshushima T, Kakinami Y. Infrasound array observations in the Lützow-Holm Bay region, East Antarctica. Polar Science. 2015;**9**:35-50. DOI: 10.1016/j.polar.2014.07.005

[11] Murayama T, Kanao M, Yamamoto M-Y, Ishihara Y. Infrasound signals and their source location inferred from array deployment in the Lützow-Holm Bay region, East Antarctica: 2015 January–June. International Journal of Geosciences. 2017;**8**:181-188. DOI: 10.4236/ijg.2017.82007

[12] Murayama T, Kanao M, Yamamoto M-Y, Ishihara Y, Matsushima T, Kakinami Y, et al. Time-space variations of infrasound sources related to environmental dynamics around the Lützow-Holm Bay, East Antarctica. Polar Science. 2017;**14**:39-48. DOI: 10.1016/j.polar.2017.10.001

[13] Walker KT, Hedlin MAH. A review of wind-noise reduction methodologies. In: Pichon AL, Blanc E, Hauchecorne A, editors. Infrasound Monitoring for Atmospheric Studies. Springer Science + Business Media B.V. Chapter 5; 2010. pp. 141-182. DOI: 10.1007/978-1-4020-9508-5

[14] Hedlin M, Alcoverro B. The use of impedance matching capillaries for reducing resonance in rosette infrasonic spatial filters. The Journal of the Acoustical Society of America. 2005;**117**:1880-1888

[15] Cansi Y. An automatic seismic event processing for detection and location:

The P.M.C.C. method. Geophysical Research Letters. 1995;**22**:1021-1024

[16] Cansi Y, Klinger Y. An automated data processing method for mini-arrays. CSEM/EMSC European-Mediterranean seismological centre. The News Letter. 1997;**11**:1021-1024

[17] Kanao M, Maggi A, Ishihara Y, Yamamoto M-Y, Nawa K, Yamada A, et al. Interaction on seismic waves between atmosphere-ocean-cryosphere and geosphere in polar region. In: Kanao M, Takenaka H, Murai Y, Matsushima J, Toyokuni G, editors. Seismic Waves—Research and Analysis. ISBN 978-953-307-944-8. Rijeka, Croatia: InTech Publisher; 2012. pp. 1-20

Chapter 8

Human Beings in Antarctica

Giichiro Ohno, Shinji Otani and Atsushi Ikeda

Abstract

Research on Antarctica has been continuing for over a century. While living in Antarctica remains difficult owing to the extreme conditions there, expeditions have progressed greatly in improving accommodations. Expeditioners are exposed to a harsh natural environment such as coldness, dryness, dramatic change in sunshine time, ultraviolet rays, and high altitude. They also live in an extreme condition: closed small groups, absolute isolation, limited equipment and supplies, and no evacuation. As such, expedition members are placed in an extreme physical and mental state. Antarctic doctors are responsible for protecting the health of members who are wintering-over. Statistical analysis of diseases showed that the most common cases were of injuries followed by internal medicine and dental problems. Some diseases were related to environmental factors. Medical operations such as medical screening expedition, remote medical care, and telehealth care contribute to the safety, and better health management systems are themselves subjects of research. Medical researches and operations are advancing and supporting one another. As a simulation of space, the Antarctic experience and the related breakthrough are utilised in space research. Outcomes of research on Antarctica contribute to the better understanding of human society as well.

Keywords: Medical research in Antarctica, Antarctic medicine, extreme medicine, remote medicine, telehealth, isolation

1. Introduction: departure for Antarctica

With a grand send off, the Japanese ice breaker *Shirase* leaves Tokyo's pier bound for Antarctica on a mid-November day. This fourth generation icebreaker is used by the Japanese Antarctic Research Expedition (JARE), a national Antarctic program established 60 years ago.

The ship heads south through the Pacific Ocean, crossing the equator, calls at Fremantle in the southwest of Australia, and loads supplies and the rest of its crews by airlift. Finally, it is time to head to Antarctica (**Figure 1**).

The waves in the Indian Ocean are rough. The ship enters the windstorm zone, tilting 60° to the left and right, making it hard for the crew to hold. More and more expedition members find that they cannot leave their rooms because of seasickness. Severe seasickness is a particular problem because it can exhaust members before arriving at Antarctica, lessening their ability to cope with the challenges ahead. To address this issue, a seasickness survey was conducted on a ship during a Southern Ocean storm in 2012. In the group that did not experience seasickness, there was no excessive decline in exhaled CO_2, which suggested that they were able to breathe slowly and deeply [1].

IntechOpen

Figure 1.
Icebreaker Shirase *en route to its destination though the sea ice. (Copy right NIPR).*

The first role of doctors going to Antarctica is to provide practical medical care. However, as this seasickness issue shows, the medical data also highlight the unique influence of the special environment of the Antarctic on the minds and bodies of human beings. The effects of Antarctica on human residing there will be illustrated in this chapter.

After clearing the storm area, the *Shirase*'s crew discovers a calm sea dotted with huge icebergs and covered with ice and snow. The ship makes its way to its destination, slowed by the need to make hundreds of back-and-forth manoeuvres to cut through the thick ice (**Figure 2**). Temperatures here are much lower than those in late autumn in Japan. Nights gradually become shorter, indicating that summer is starting in Antarctica. During the "white night", when the sun does not set, the crew gathers on the bow deck to see with their own eyes the immensity of the polar continent.

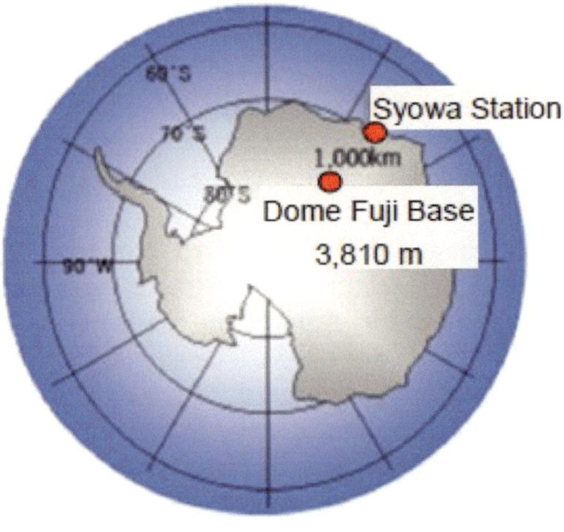

Figure 2.
Syowa Station and Dome Fuji Base. The Japan Antarctic Expedition team built Syowa Station (S 69°00′, E 39°35′) in 1956 and has continuously conducted wintering activities ever since. In 1995, Dome Fuji Base (S 77°19′, E 39°42′), altitude of 3810 m) was built inland about 1000 km away from Syowa Station.

One month after leaving Japan, they arrive at Antarctica 15,000 km away from home. This is not the end of a long journey rather the beginning of a year of surprising mysteries.

2. Antarctic station: entrance to the unknown world

Syowa Station, the Japanese Antarctic station, was built in 1956 on Ongul Island (S 69°0′, E 39°35′) 4 km off the main landmass. One of the key features of the polar region is the dramatic difference in night-time: there is both a polar night period (24 hour of darkness) and a white night period in which the sun shines all day. At Syowa Station, from November 20 to January 20, the sun does not set, and from the end of May to July 10, the sun never rises.

The eastern part of Antarctica, where the station is situated, is known for its inaccessibility. Because Japan started to explore Antarctica later than other developed nations, it had no choice but to build its station on this isolated island (**Figure 3**).

Each summer, the icebreaker carries not only a new wintering participant but also the fuel, food, and observation instruments needed for overwintering. When the short summer is over, the ship must leave the shores of Antarctica for Japan lest it be stranded in the ice, carrying with it the researchers and logistical staff whose annual mission has come to an end.

In winter, members of the station are truly isolated; the ship cannot keep pace with the thick ice mass. To complicate matters further, air routes are closed by stormy weather conditions and extreme darkness. This makes it impossible to deliver additional supplies or expedition members. Antarctica becomes almost lifeless and impenetrable. Even cold-resistant penguins and seals disappear from the landscape. The wintering-over team at the outpost must live in the narrow, enclosed space without means of escape for a long year.

Station members endure harsh conditions caused by both nature and the confinement of their outpost; indeed, the conditions are similar to what might be experienced on a Mars station in the future.

Three factors determine the number of wintering team members (**Figure 4**). One is the amount of goods that can be carried to the station, which has increased as the new generation of ship enters service. The immense thickness of winter ice makes it impossible for even a modern ship to reach the station. However,

Figure 3.
Syowa Station. The station is located on Ongul Island, 4 km off the Antarctic continent. There are no nearby bases, and when a ship departs for Japan, the station is completely isolated for 1 year.

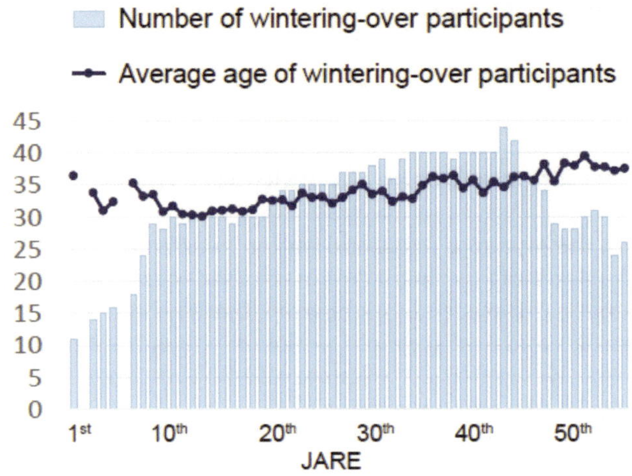

Figure 4.
Trends in number of crew members and their average ages. The first team was composed of 11 members. The number has increased to peak at 44 and then decreased to around 30. The average age has increased gradually. The most recent team's average age is 37 years.

icebreaking capacity has progressed significantly since Japan began sending Antarctic expeditions, and Japan has been forced to abandon overwintering of the second expedition team in 1957 and to reduce the number of wintering-over members for the 55th and-56th expedition in 2013—2014 when only one doctor was available. The second factor affecting wintering team member size is the capacity of the station. The station is expanded every year, and it is scalable to accommodate more than 40 wintering members. The third is the size of the research plan. Due to observation automation and the use of aircraft in the summer, the number of summer parties expanding their activities has increased and the number of wintering members is decreasing. Japan's first team consisted in 11 members; the number increased, peaking with 44 members in 2003. More recently, wintering parties have comprised about 30 people.

The wintering-over member's average age has been trending upward year and year. In 2015, the average age of wintering team was 37 years old.

Women have been participating in Japanese Antarctic wintering-over team since 1998. They totalled 5% of participants from 2002 to 2015.

The composition of the wintering-over party is decided a year prior to departure. Candidates are subjected to a rigorous health check including head CT, gastroscopy, abdominal ultrasonic examination, and electroencephalogram. The examination also includes a psychological assessment. Researchers and some logistics staff are selected by recommendation. Other staff members, such as medical doctors and cooks, are recruited through public offering.

Each new group meets when they begin to prepare for the trip. The selected candidates participate in two training camps in winter and summer. After completing the camp trainings, they start gathering supplies at the expedition centre 5 months before departure. They also complete a 24-hour shared life experience on the outbound ship. Through this training, mutual understanding and relationships gradually develop. In such isolated conditions as those in Antarctica, cooperation within the team is indispensable not only for research but also for the safety of the station and wintering-over members. Despite their extensive training, some members might have difficulty adapting to their new environment. Cases of severe depression and suicide have been reported at other international stations. Methods for screening for unsuitability at the time of selection have been studied, though this remains a significant challenge, as many people who manifest psychological and social problems during overwintering are known to have led a normal life before setting foot in Antarctica. The behavioural gap makes it difficult to identify individuals who may be unfit for the mission during the selection process. A screening method that can be universally recommended has not yet been found.

Meanwhile, the importance of leadership in the Antarctic context is of growing interest. Spending a long time in isolation is always challenging. Managing individual personalities for the benefit of the team is critical to the success of the mission and the wellbeing of all members.

3. At the bottom of the dancing aurora

White nights end around mid-January. Towards the end of February, nights begin to become darker still, and during nice weather, the aurora dazzles onlookers in against a sky full of stars. The length of the day decreases by 5 hours per month up to the polar night. The speed of change is fast enough for observers to notice the difference in brightness within 2 days. The weather deteriorates and the temperature drops rapidly.

At the outpost, work progresses smoothly. The medical office has adequate space in the narrow station. One of the two station doctors is a surgeon. Most stations in Antarctica have only one doctor, which is risky, as he or she may also require medical attention. To avoid this hazard, stations such as JARE make the presence of two doctors mandatory. There are no other medical staff such as nurses.

The total number of injuries and diseases during JARE's first 43 wintering-over periods is 4931 cases (from 1956 to 2002) [2]. The percentage of occurrences sorted by specialty is shown in **Figure 5** above. The most common cases comprise injuries, followed by internal medicine issues and dental problems, respectively. Dental problems are known to be prevalent at every Antarctic station. Because all kinds of medical troubles including dental problems occur among Antarctic overwintering participants, doctors universally act as generalists.

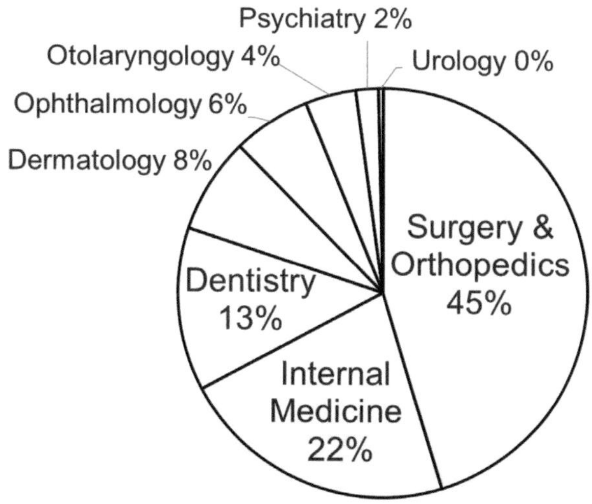

Figure 5.
Medical consultations of wintering crew members. The total number of medical consultations at Syowa Station cases from 1956 to 2003 was 4931. Among them, 4633 cases with known onset month were analysed. The most common cases were surgery and orthopaedics, followed by internal medicine and dental problems, respectively. The diseases covered every field.

The sole death at Syowa Station was caused by distress in a blizzard. The majority of the recorded 76 deaths at Antarctic stations were caused by accidents (e.g., aircraft accidents, falls, distress, and carbon monoxide poisoning). However, acute cardiac insufficiency, such as a myocardial infarction, has also claimed wintering-over persons [3, 4].

The medical post in Syowa Station is equipped much like a small hospital in Japan: it has the capacity to conduct X-ray fluoroscopy, electrocardiograms, blood tests, and abdominal ultrasounds in addition to general examinations [3, 4]. Although the equipment is rarely used, it is always maintained as well as in Japan so as to handle any emergency; the station is prepared to treat severe cases, all anticipated diseases expected to occur, and even rare but serious infectious diseases such as tuberculosis.

Syowa Station has an operating room with enough equipment to conduct surgical operations under general anaesthesia. To date, two appendectomies have been conducted at the station. Other stations have reported operations relating to appendicitis, subdural hematoma, and fractures. Surgical operations are extremely difficult to perform in Antarctic conditions for a variety of reasons, such as difficulty maintaining anaesthesia, applying blood transfusion via "walking blood bank", preparing an aseptic operation space and working without assistants such as nurses. In a case at a Russian station in 1961 illustrating the risks of relying on a single physician, the station's sole doctor was forced to perform surgery on himself when he suffered from appendicitis [5].

Given capacity limitations, there are cases when evacuation to Japan becomes necessary. At Syowa Station, four medical evacuations have been undertaken for patients who suffered pelvic and femoral fractures, acute renal failure, and arrhythmia. Three cases required sea evacuations that lasted 25 to 120 days. The other case was air lifted to Japan. This patient was carried by a small regional plane to Norway's Antarctic station 1200 km away, and this is one of the few stations equipped with a runway that an intercontinental airplane can use. The patient was then transported to Japan via South Africa over a period of 6 days. Although the use of an aircraft drastically shortened the evacuation duration, it cannot be said that the procedure is fast enough to suffice in case of a severe emergency. It was fortunate that all evacuations happened in summer, when it was still possible to transport patients off the continent [3, 4].

Every country operating in the Antarctic, including Japan, pursues the possibility of an evacuation route, especially in winter. Though a few intercontinental air operations have been conducted in winter as unusual events at certain stations under extremely favorable conditions, it is nearly impossible for any evacuation or rescue team to operate either by sea or air during winter because of limited visibility and bad weather.

Medical collaboration occurs between various nations in areas where stations are adjacent. There are nine stations of representing eight countries on King George Island at the northern tip of the Antarctic Peninsula. In these areas, mutual medical cooperation is advancing through consultation on specialised fields, shared use of inspection equipment, and so on. However, most of the stations on the continent, including Syowa Station, are isolated and do not have any nearby stations with which to cooperate.

In such situations, telemedicine plays a vital role; it has been successfully operationalised in Antarctica. Nonetheless, it is notably restricted by the technology available. In 2004, a satellite television connection was installed, linking Syowa Station and domestic hospitals in Japan (**Figure 6**). Telemedicine using this system is highly effective for both diagnosis and treatment in orthopaedic medicine (e.g., fractures, arthritis, and ligament injuries), rehabilitation, and dentistry, among

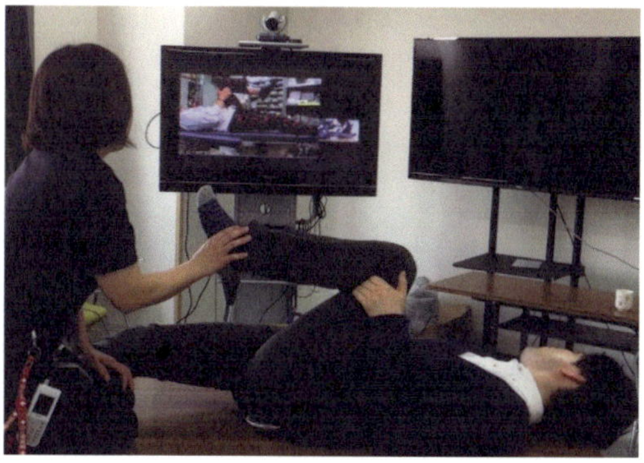

Figure 6.
Telemedicine Scene. The patient has a knee ligament injury. To evaluate the joint motion and pain, the therapist in Japan (left side in foreground) demonstrates the procedure on the simulated patient (right side). Then, Antarctic doctors (on the screen) perform as instructed by the therapist in Japan, who diagnoses the case and recommends a therapy plan.

other fields. However, in certain areas as audiology and dermatology, the technology shows some limitations because of the difficulty in taking images or ensuring a high enough resolution [6]. In summary, consultations with colleagues in Japan have advantages for both patients and Antarctic doctors who deal with patients issues outside their own specialty and within the restrictions of the limited local facilities.

An Antarctic doctor lives with several constant stresses, including the tension between the knowledge that their services may not be required—healthy young participants rarely need doctors' aid—but , once they are needed, that need may be extraordinary, and they may not be able to save the lives of those suffering from severe issues.

4. Careless frostbite and unpreventable frostbite

It is the first experience for almost all doctors to visit Antarctica. They are confused by the difference of medicine between Antarctic environment and the daily site. However, if it is possible to predict the diseases which may occur and the time of onset, doctors can prepare themselves and take preventive measures.

For that purpose, an analysis was done on the medical consultations of participants at Syowa Station during the first to 43rd wintering over (from 1956 to 2003). The total number of wintering-over participants was 1278; 4931 medical cases occurred during this period. Among them, 4633 cases in which the month of onset was clear were used to analyse rates of incidence by month.

In the analysis of each year, there are limitations in seeing the original tendency because the number of cases is small, and circumstances unique to each year may exert influence. Aggregation over 40 years eliminates these problems.

Figure 7 shows numbers of monthly consultations. The cases increase at the start of overwintering and sharply drop during the polar night. Thereafter, the rate shows a gradual decreasing trend with no particularly notable changes.

Frostbite increases and peaks in May (**Figure 8**). This coincides with late autumn which is not the coldest season of Antarctica, experiencing temperatures of −15 to −20°C (rarely experienced in Japan) that drop rapidly. Such an increase in frostbite incidence may be due to carelessness and unpreparedness.

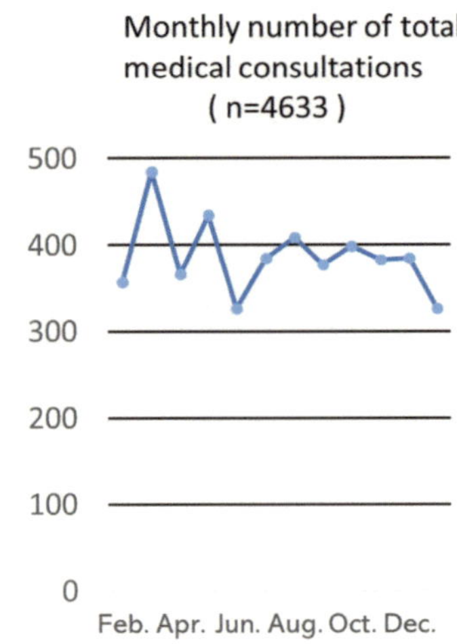

Figure 7.
Monthly evolution of all 4633 cases recorded from 1956 to 2003. The total number of medical consultations at Syowa Station was 4931 cases from 1956 to 2003. Among them, 4633 cases of with known onset month were analysed. There is no particular seasonal variation, but the cases frequently occur at the start of overwintering, and there is a gradual decreasing trend thereafter.

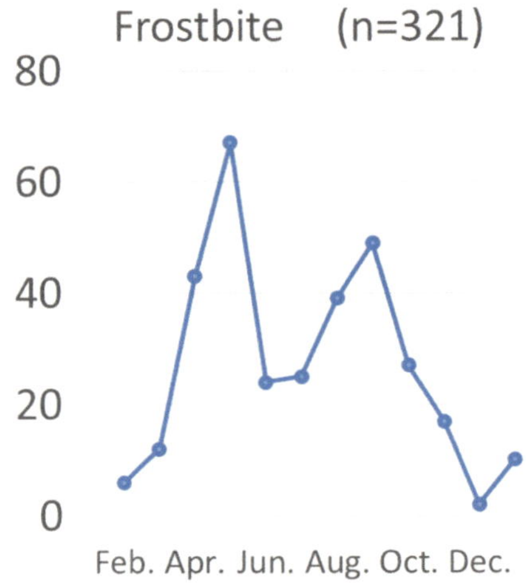

Figure 8.
Monthly occurrences of frostbite. A total of 321 cases of frostbite occurred from 1956 to 1999. In the aggregate, incidence peaks twice: the highest in May and the second in September, during the coldest season.

Subsequently, in the polar night season, frostbite cases decrease because outdoor work is restricted.

August and September are the coldest months with temperatures of nearly −40°C. In addition, outdoor activities come into full swing. Frostbite case increases again, reaching a second lower peak that does not exceed the first. Frostbite is considered unpreventable because these cases still occur despite awareness policies and preventive measures.

5. Viruses in closed populations always require the next new population to survive

According to the same aggregate data, respiratory diseases totalled 218 cases, including instances of the common cold and upper respiratory inflammation

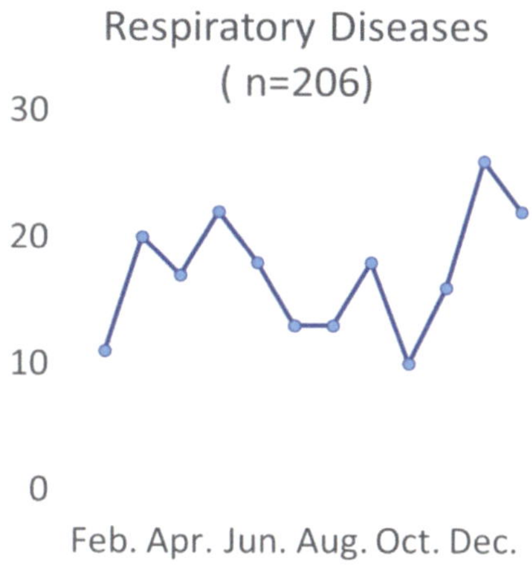

Figure 9.
Monthly occurrence of respiratory diseases. There were 218 cases of respiratory diseases from 1956 to 1999. They tend to occur at the beginning of overwintering and then decrease gradually before soaring in December.

(**Figure 9**). Environmental factors such as coldness and dryness contribute to these diseases, and the viruses that are the biggest causes of these illnesses in Japan may also be the culprits here. However, there are no viruses which naturally occur in the harsh environment of Antarctica, and it is hard for viruses brought in by humans to survive among the small isolated group of station occupants once the people

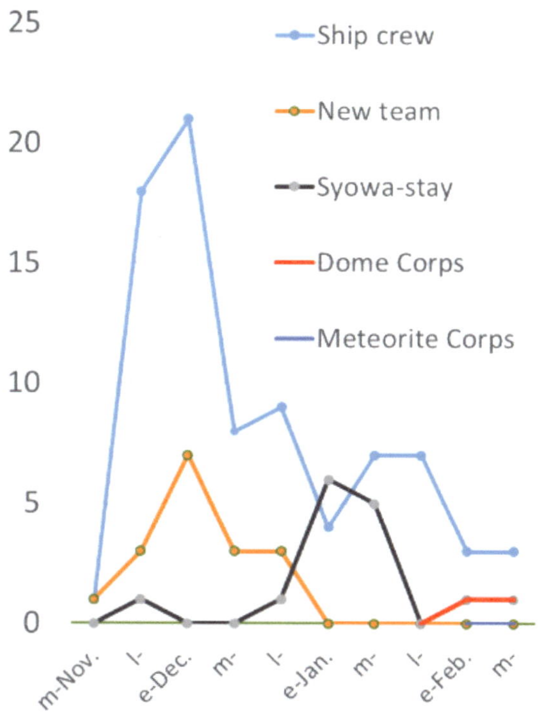

Figure 10.
A cold outbreak on the Icebreaker ship and at Syowa Station after the ship's arrival. The number of newly cold-affected people is shown by period: early (e-), middle (m-), and late (l-) in each month. Groups include "ship crews" and "JARE 40", who were new expedition members arriving from Japan on the ship as part of the 40th expedition. Members of the JARE 39th expedition were divided into three groups when the ship arrived at the station: 25 members were at the station; 6 members were at Dome Fuji base, returning to the station back in January; and 8 members were absent on meteorite survey inland and came back in February.

infected have acquired immunity. Looking at the disease occurrences by month, it is shown that respiratory ailments are frequent at the beginning of overwintering and then decrease gradually until soaring in prevalence in December. It is commonly known among Antarctic excursion members that a cold outbreak will present as soon as a ship has come for the first time in a year.

In one example, when the ship arrived at the station where the participants of the 39th season were finishing their 1-year term, 25 members stayed on at Syowa Station. Six members of Dome F corps and eight members of the meteorite corps were absent when the ship arrived.

A cold epidemic occurred among the outbound ship crews and new expeditioners on the ship (**Figure 10**). The outbreak was dying down when the ship arrived at Syowa Station. Immediately after the ship's arrival, an outbreak of colds occurred among the station members. When the Dome F team came back to the station, the outbreak had almost faded but two of six members caught a cold. Eight members of the meteorite corps came back to the station, 1 month after the arrival of the ship, and no cold case was recorded. This is an extremely interesting demonstration of virus behaviour in a closed group.

6. Diseases and daily living factors

There were 199 documented injuries and diseases in 1 year among the 39th team, which had 39 participants. The causes of all cases were investigated.

Diseases and injuries that occurred during overwintering are classified into three categories: health problems attributable to the natural environment (63 cases), cases related to indoor and outdoor work (55 cases), and cases related to daily life other than tasks (34 cases). The monthly number of these three categories is shown in **Figure 11**.

Natural environment-related diseases are frequent in the coldest season. Work-related diseases are high at the beginning of overwintering and from spring to summer, when outdoor activities are resumed. Lifestyle-related issues occurred throughout the year, with a tendency to increase at the beginning and end of overwintering.

Human factors may also be at play. Looking at the number of medical cases per capita among the team, there was no difference between researchers and logistical members. However, a clear difference was shown among generations; the morbidity rate is lowest in the 40s and higher among members in their 20s and 50s (**Figure 12**) [7]. This suggests that life experience might be involved in the

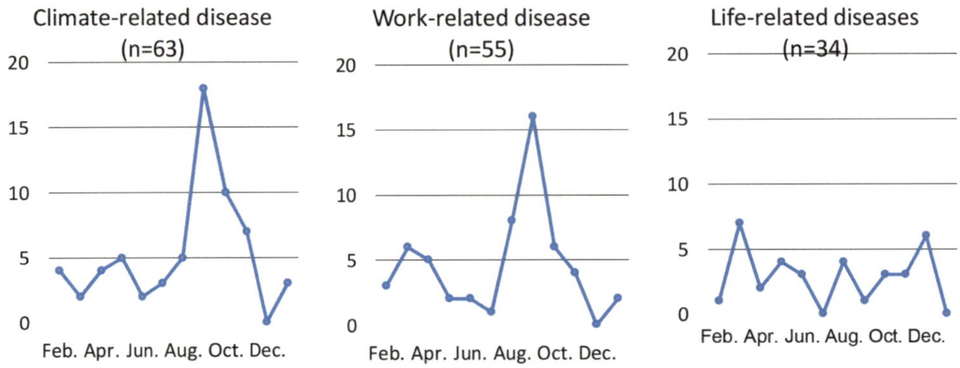

Figure 11.
Monthly change in number of diseases attributed to natural environment, work and ordinary life. Among 199 cases that occurred during one year at Syowa Station from February 1998 to January 1999, 63 cases were due to natural environment, 55 cases to work, and 34 cases to daily life. There was a seasonal change in occurrence in each disease category.

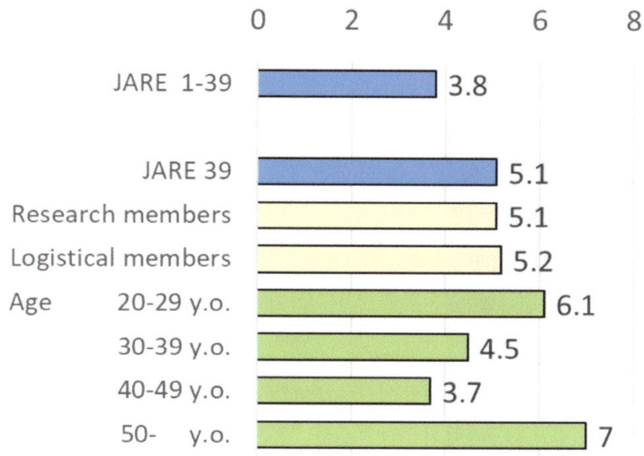

Figure 12.
The number of medical cases per capita on the 39th wintering-over team. The wintering team of JARE 39th was composed of 39 participants and had 199 medical consultations. Frequency of injury and disease was compared between research and logistical members, between generations. (Quoted from reference [7] and partially modified.)

occurrence of injuries and diseases in addition to pointing out the relevance of physical age in extreme environments.

7. Bond strengthening at Antarctic stations

At Syowa Station, there are up to 10 activity teams for entertainment, farming, sports, movies, drinking, and more. Members are responsible for several tasks beyond their official work. These activities are thought to create an atmosphere promoting strong and smooth relationships at the station to prevent loneliness.

The sports team organises monthly events such as baseball, golf, and dodge-ball matches. Golfing in the snow has resulted in many lost balls. A beautiful game of football was once played on sea ice with diamond dust sparkling in the sky. However, few members are familiar with playing sports on the snow and ice; as a result, sports activities cause 10% of medical consultations during overwintering. Despite the potential for injuries, sports are believed to contribute to the prevention of severe medical cases by allowing members to experience and enjoy the Antarctic environment.

The entertainment team plans season-based events in monotonous Antarctica. For instance, in July, *Tanabata*, a Japanese traditional star festival, is celebrated. Everyone hangs two wish cards on a bamboo-like branch. Wishes recorded on the cards were as follows: 29% of participants wished for “Safety & Health”, 26% for “Something not in Antarctica”, 21% for “Family”, 16% each for “Work”, “Lover”, and “Weight loss”. The wishes are considered to be different expressions of experienced stressors. This result is similar to finding from a study on single Japanese individuals living overseas.

8. Losing the sun

Entering the overwintering period, the day, ever shortening, finally disappears. Losing sunshine for a complete day is characterised by the experience of darkness, coldness, and a sense of the solar loss. People feel sleepy at noon and have difficulty sleeping at night. Confusion abounds; when a person wakes up and sees the clock showing 1 o’clock in the dark, he or she cannot judge quickly whether it is daytime or the middle of the night.

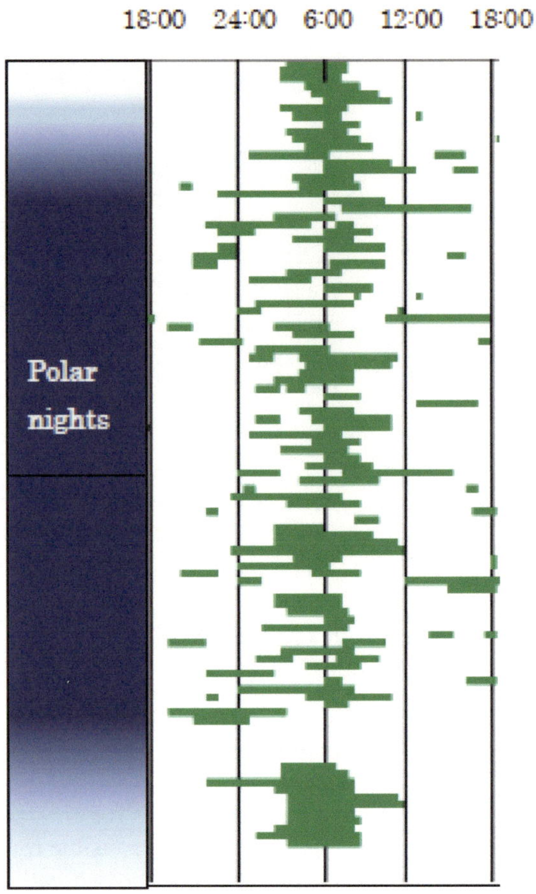

Figure 13.
Trend of sleeping hours of one crew member. The vertical axis shows the period from May to July, while the horizontal axis depicts the time of day; the green belt is the sleeping time. The person is a male mechanical member in his twenties. He originally had a habit of staying up late. Regularity of sleeping hours disappears during the polar night period. After the polar night is over, his work-load increases, which improves sleeping patterns.

The human biorhythm is said to consist of a 25-hour cycle, which is 1 hour longer than a normal day. This cycle is reset by the brightness of morning sunshine.

During polar nights, the sun never appears during the day. The human biorhythm will not be reset and will fall into its 25-hour cycle. As social life is another trigger to reset this rhythm apart from light, the regimented timetable of station life ensures that the rhythm sustains the typical 24-hour cycle.

However, during this period, outdoor activities cannot be carried out because of the darkness, a measure to prevent accidents. Monotonous days become the norm and are less effective in resetting the circadian rhythm (**Figure 13**).

In addition to sleep disturbances, the darkness, cold, and confined nature of station life also make people feel depressed or uneasy.

9. Midwinter

Surviving this dark period safely is one of the biggest goals of station management. Various innovative activities have been initiated to address the issue. One of them is the midwinter festival.

The summer solstice of the northern hemisphere in late June coincides with the midwinter day, which is celebrated grandly in Antarctica.

The festival consists of certain special events and a grand dinner. Anticipation of this celebration, and preparation for the events and dinner, gives the station a bright atmosphere for several days.

This festival has been held since the time of Amundsen and Scott, competitive pioneers who each sought to be the first to reach the South Pole. Even now, the midwinter festival is held at most Antarctic stations, which send congratulatory messages to one another.

After the festival, the participants plod through the other half of the polar night season, and the blurred line between reality and dreams stretches on.

In mid-July, people gather at a small hill behind the station. At noon, the sun shines over the side of an iceberg for a moment before sinking behind another one. The polar night season is over. Although the end of darkness is only a short moment, when the sun peeks all too briefly out to turn the blue world to an orange shine, each member of the crew witnesses the scene and feels the moment intensely.

Despite the sun's gradual return, the cold becomes increasingly severe, yet outdoor operations increase with the daylight. Each person has clear tasks. During this period, sleep quality and rhythm improve.

10. Physiological changes in the Antarctic inland

Antarctica is covered by ice that averages 2000 m in thickness. Therefore, the Antarctic inland is an extremely cold place, and the air is rarefied. Dome Fuji Base (Dome F), an inland Antarctic research base operated by Japan, is located at S77°19′, E39°42′. It is 3810 m above sea level and separated from Syowa Station by about 1000 km (**Figure 14**). The annual mean temperature at Dome F is −54.3°C; the lowest temperature (−79.7°C) was recorded in 1996.

In general, an inland JARE party typically comprising around 10 members leaves Syowa Station with several large snow vehicles on a 4-week drives to Dome F. Extreme environmental conditions in the Antarctic inland may cause hypoxia, mountain sickness, frostbite, depression, and so on. To better protect the safety of excursion members during Antarctic inland expeditions, we are collecting physiological data from members of inland parties.

Figure 14.
Dome Fuji Base. It is located at S 77°, E 39°, 3810 m above sea level and separated from Syowa Station by about 1000 km, where the snow field extends as far as the eye can see. Most of the base facilities are under the snow.

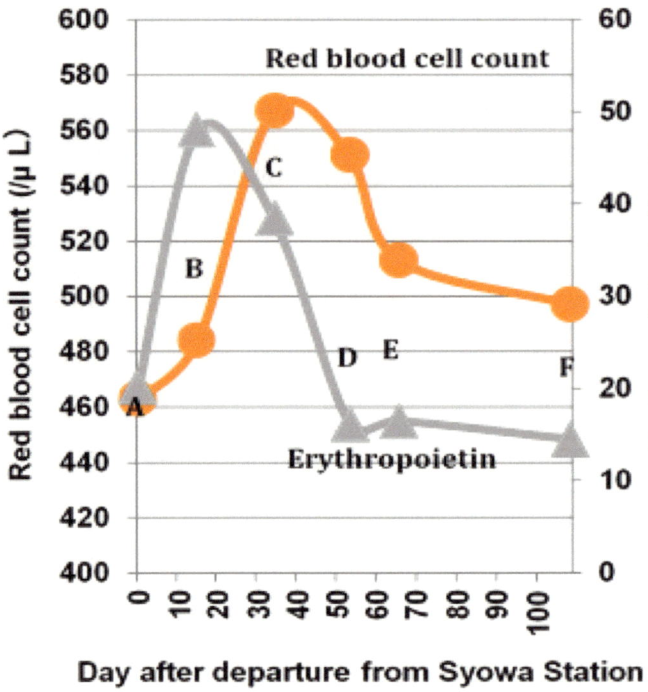

Figure 15.
Trends in the serum concentration of erythropoietin and red blood cell count. Sampling point: A, Syowa Station (29 m); B, 3032 m; C, Dome F; D, 2960 m; E 2077 m; F, Syowa Station.

Among party members, the serum concentration of erythropoietin quickly increased, subsequently leading to increased red blood cell count in a seven-member party which sets out during the austral summer of 1999–2000 (**Figure 15**). Haematological adaptation to a high altitude was completed in several weeks [8]. Percutaneous arterial blood oxygen saturation (SpO_2) levels from four parties (totalling 33 members) had a significant negative correlation with higher elevation above sea levels. However, after arriving at Dome F, party members` SpO_2 levels showed a positive correlation in the duration of their stay. This may be explained by effective high-altitude adaptation [9] (**Figure 16**).

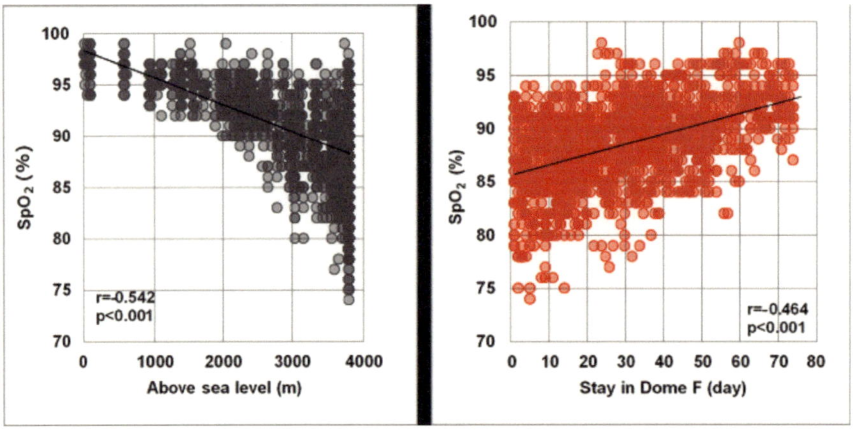

Figure 16.
Correlation between percutaneous arterial blood oxygen saturation (SpO2) levels and altitude above sea level (left), and duration of stay (days) at Dome F (right).

None of the participants suffered from severe mountain sicknesses like pulmonary or brain oedema, although some experienced headache, fatigue, sleep disturbance, or elevated blood pressure because of the slow increase in altitude. Most of party members remained in relatively good health during their Dome F stay. This demonstrates that humans can successfully survive in and acclimate to the extreme environmental conditions in the Antarctic inland.

11. Human being as an indicator of Ozone hole

After the polar night, daytime becomes longer by 5 hours a month, leading into the white night season. There are four clear seasons with noticeable light changes in Antarctica.

In October, the sun reflects on brand-new, pure white snow and ice, creating a world of light. During this time, the ozone hole is emphasised.

The ozone hole was first observed in 1980, and it continues to expand.

Ultraviolet rays from space are pouring through the hole in the ozone layer, which normally blocks them.

We have compared the ultraviolet rays in Antarctica in October with those at the equator (**Figure 17**) [10].

The results show that ultraviolet rays emanating from the sun's precise direction are as strong as those at the equator, and that the ultraviolet rays irradiating a person's face are stronger in Antarctica because of the low sun and reflection by the snow and ice (**Figure 18**).

Diseases caused by ultraviolet rays include dermatitis, conjunctivitis, and facial dermatitis. The incidence of these issues is low in winter and tends to increase in summer, which shows a correlation with the quantity of ultraviolet rays encountered. The highest incidence spans from October to December. Looking at the number of occurrences at this time in terms of annual change, a trend of gradual

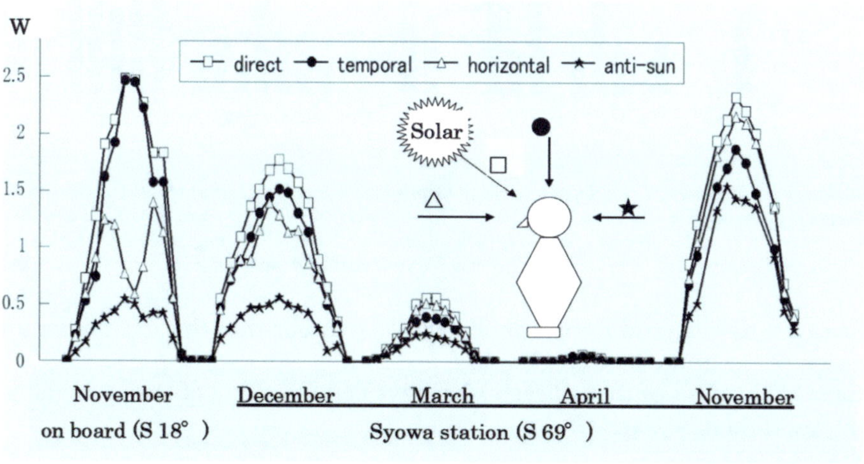

Figure 17.
Change in UVB radiation dose by latitude, season, direction, and time of day. Measurement of the ultraviolet ray radiation doses was conducted at S 18°on November 25, when the sun is directly overhead, and at Syowa Station in December (during the white nights), March, April, and November (under the ozone hole). The measurement was made every hour from 7:00 to 19:00 in four directions (zenith, direct sunlight direction, horizontal direction of the sun's direction, and horizontal direction opposite to the sun). Ultraviolet rays in Antarctica in November were as strong as those at the equator. Ultraviolet rays irradiating the face of a person were stronger in Antarctica because of the low sun and the reflection by the snow and ice. (Quoted from reference [9] and partially modified.)

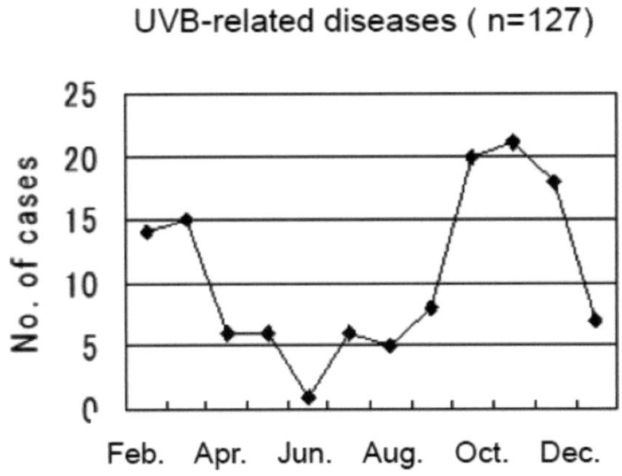

Figure 18.
Monthly occurrence of UVB-related diseases. A total of 127 cases of UVB-related diseases were recorded from 1956 to 1999. Here, dermatitis, conjunctivitis, and facial dermatitis are compiled as UV-related diseases. Not all of these are exclusively caused by ultraviolet rays; for example, they may be affected by the amount of outdoor work. The incidence is low in winter and tends to increase in summer, which shows correlation with the amount of ultraviolet rays. The highest peak is recorded from October to December, when the ozone hole appears. (Quoted from reference [9] and partially modified.)

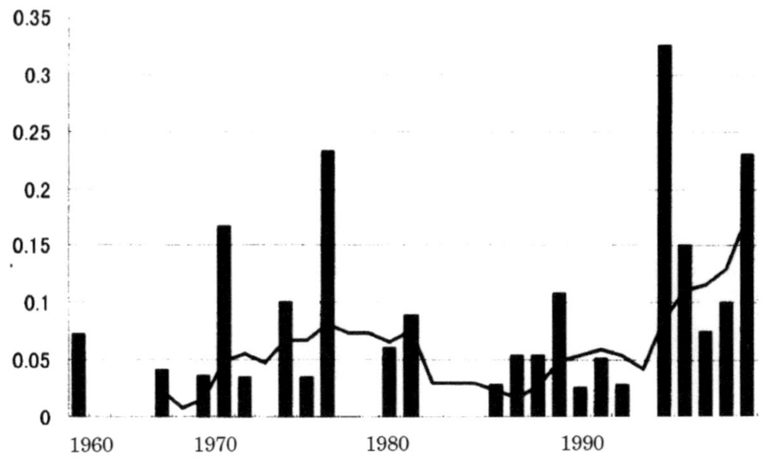

Figure 19.
Annual change in incidence of UV-related cases in the ozone hole season. Time series of total number of UV-related cases per person from August to December and moving average from 1956 to 1999. The ozone hole first appeared in the 1980' and continues to expand. Both annual incidence and moving average have been increasing since the late 1980s (Quoted from reference [9] and partially modified.).

increases can be observed from around 1980, consistent with the appearance and expansion of the ozone hole (**Figure 19**).

Statistical analysis of these diseases has revealed the effect of the ozone hole on human health in this context.

12. Third-quarter syndrome

For many team members, wintering-over is a year in which they will endure stresses that they never previously experienced.

Diseases attributed to stress include insomnia, headaches, urticaria, hypertension, and digestive troubles. The annual occurrence of such diseases shows a

DOI: http://dx.doi.org/10.5772/intechopen.81974

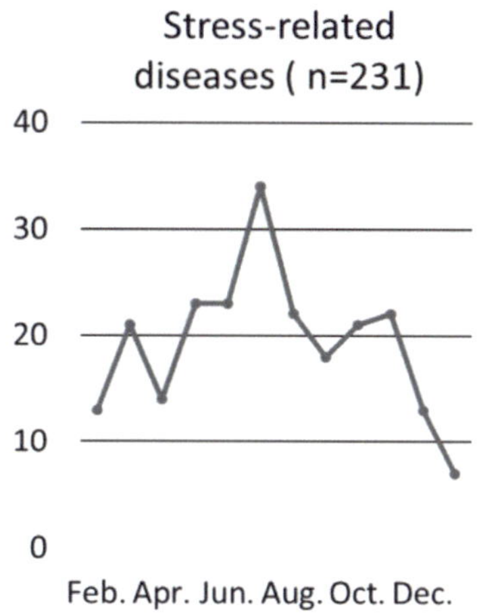

Figure 20.
Monthly change in number of diseases attributed to stress. A total of 231 stress- related diseases were recorded from 1956 to 1999. A small peak is observed at the beginning of the wintering-over period, increasing continuously until the polar night period, when it makes the highest peak. Subsequently, it decreases gradually. In early summer, incidence increases again until reaching a second lower peak.

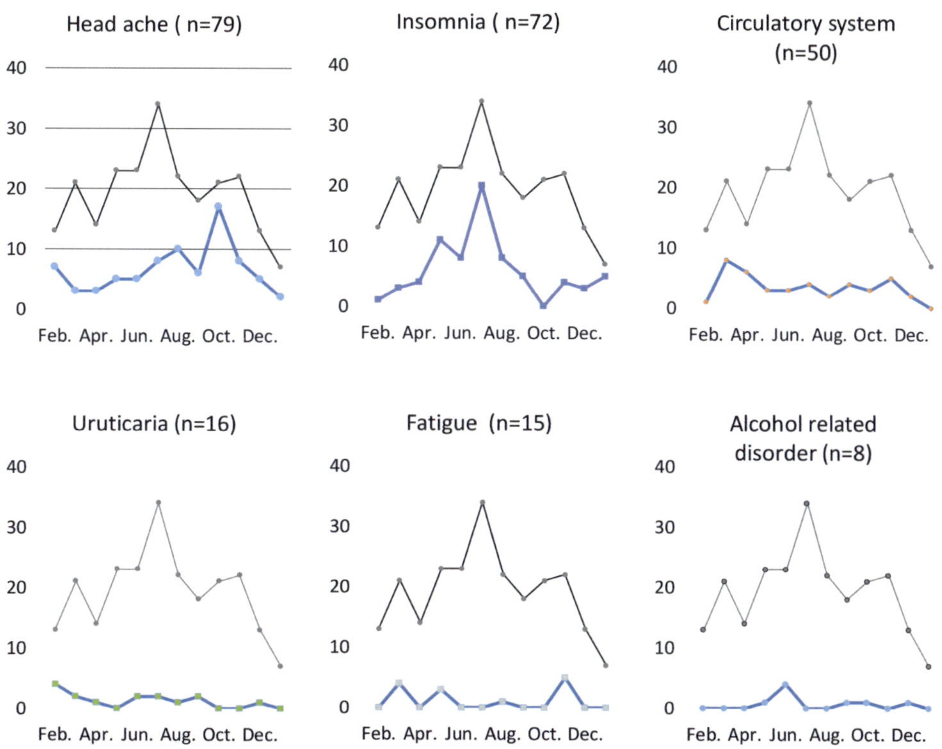

Figure 21.
Changes in numbers of major diseases attributed to stress. Several diseases caused by stress, including headaches, insomnia, circulatory system issues, urticaria, fatigue, and alcohol-related disorders were indicated in 228 cases at Syowa Station from 1956 to 1999. The blue line shows each disease, and the black line in each graph shows the total number of stress-related diseases. Each disease showed a characteristic distribution. The peak of all stress-related diseases is the sum of these various diseases, that is, various stresses.

three-peak pattern, highest at the beginning of overwintering, during polar nights and in late spring (**Figure 20**). The last peak coincides with the so-called third-quarter syndrome. It is hypothesised that the third quarter of the mission is when psychological crises tend to arise, such as a drop in morale and a rise in negative emotions.

In a psychological study of 10 consecutive years from 2003 to 2013, such a phenomenon was recognised in seven parties. Various fluctuations were found in the number of injuries and diseases experienced by these parties. However, in other parties, the number of medical consultations decreased during this time. These facts may suggest that while the harsh nature of Antarctica is similar every year, the cause of stresses experienced by team members differs each year, perhaps in accordance with human group formation changes every year.

Looking at the number of occurrences by month of each stress-related disease individually, we observe seasonal variations peculiar to each (**Figure 21**). But these changes have no relation to a seasonal change in the total number of all stress-related diseases and there is no similarity between diseases.

This suggests that the third-quarter syndrome exists, but it is not inevitable or consistent in its emergence. Second, we may infer that stress in the Antarctic is, in fact, a complex of a wide variety of stresses. Moreover, the stress that causes stress-related diseases can differ from disease to disease.

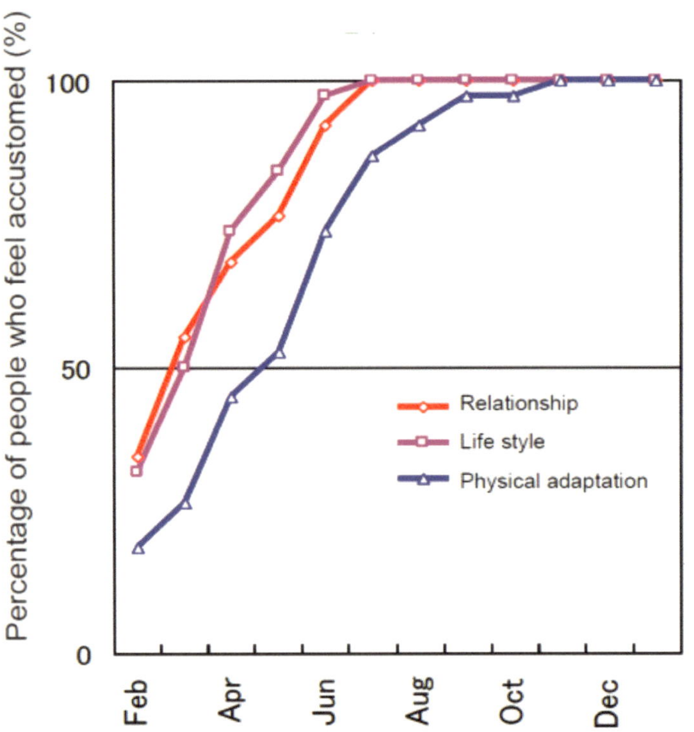

Figure 22.
Percentage of people who feel accustomed to Antarctica. The data source is a questionnaire survey on when each member felt he or she became accustomed to Antarctic relationships (orange line), lifestyle (pink) and physical condition (blue) during their overwintering year. Many persons indicated feeling that they had gotten used to Antarctic life gradually. Regarding human relations and lifestyle, all members answered that they got accustomed by July, when the polar night is over. Meanwhile, physical adaptation appears to take until November. Dramatic changes in the environment that participants had never previously experienced appear one after another. Therefore, adaptation is a continuous process that lasts until the final overwintering period.

Thus, various stresses that cause changes in each disease and the sum of those stresses are causing a yearly change such as the third-quarter syndrome.

Numerous psychological investigations have been conducted at Syowa Station [11]. Further investigation is necessary to elucidate these problems.

13. The fifth-quarter? Difficulty in acclimatising

Each December, the icebreaker arrives at Syowa Station for the first time in a year. The ship brings the next wintering-over participants wearing brand-new cold-weather gear and bearing fresh vegetables, fruits, and consigned items from the families of wintering-over members.

After finishing the handover to the new participants, the previous participants withdraw sequentially.

The helicopter circles the station before heading to the ship. The departing members recognise later that they had just taken their very last step in Antarctica.

How do wintering-over members adapt to the environment and life in Antarctica on their first visit? One-fourth of participants reported that they were able to live without feeling discomfort from the beginning. Although they are embarrassed at their lack of experience, they soon adapt. Many members have indicated that they get used to Antarctic life gradually (**Figure 22**). Regarding human relations and lifestyle, all surveyed participants answered that they got acclimatised by July when the polar night ends.

Meanwhile, full physical adaptation takes time, usually not occurring until November. Dramatic changes in this wholly novel environment appear one after another. Therefore, adaptation is a continuous process that lasts until the end of the overwintering experience.

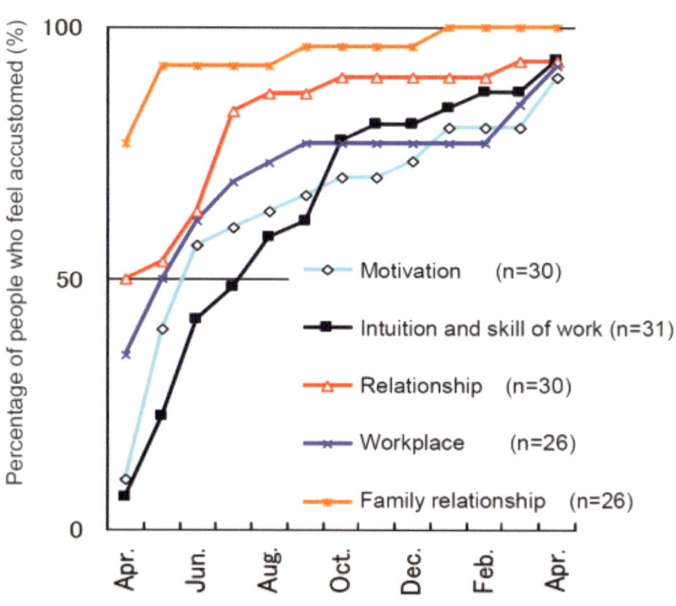

Figure 23.
Reintegration after living in Antarctica for a year. This graph shows the percentage of crew members who feel accustomed to Japan after returning home. Family relationships (orange) recover as if there was no absence. Half of participants perceive difficulties at work (blue) and with social relationships (red) upon returning. Many members recover in about 3 months, whereas some members need more than 1 year to reintegrate. Most people face difficulty at the beginning, and it takes 1 year to regain their intuition and skill at work (black). It takes time for many members to regain motivation (light blue).

Reintegration after living in Antarctica for a year is also a process of interest? Data show that family relationships recover as if there was no absence (**Figure 23**). However, half of participants reported perceiving difficulties in the workplace and with social relationships back home. Many members recover in about 3 months; those who do not recover by that point tend to remain in that tenuous state for a year. It seems to take much more time to adapt to workplace changes after returning home, being absent for a year means re-entering a workplace that may have changed procedures or staff. Most people face difficulty in the beginning, and it takes around 1 year to regain one's contextual intuition and catch up technologically. In addition, it takes time for many members to regain their motivation.

Interestingly, it seems that the strongest group—those who most easily became accustomed to the harsh Antarctic life—struggles the most to return to their lives in Japan. Human society may be harder to adapt to than the Antarctic.

14. Conclusion: departure from Antarctica

Research on Antarctica has been continuing for more than a century. The first expedition started as a challenge to survive in an unknown area. Then, the context of excursions changed to an investigation of Antarctica. At present, the purpose of research is to gain a better understanding of both earth and space.

It is still difficult to live in the extreme Antarctic climate, although expeditions have made great progress in improving the available accommodations. Expeditioners are exposed to a harsh natural environment and live in extreme conditions. The role of doctors going to Antarctica is to ensure the safe return of all overwintering members.

To this end, the effects of this dramatically different environment on the human body and mind need to be understood. Medical treatment itself in Antarctica is a subject of research, as in the case of extreme remote medicine.

Elucidation of these mechanisms and techniques is indispensable for providing medical care that protects local health regardless of environmental conditions. Furthermore, this area of medical research offers a treasure trove of interesting and varied themes for further exploration.

Antarctica has come to be studied as a simulation of space. The Antarctic environment is said to be similar to that of Mars, and an Antarctic station may be compared to a Mars station or a spaceship. Antarctic medicine is also being connected to space medicine. Thus, the value of continuing research in this environment cannot be understated.

Phenomena in Antarctica are extraordinary. What happening on human being may be emphasized and becoming noticeable, and they share commonalities with those in Japan. Experiences and breakthroughs in Antarctica will hopefully help provide both a platform for expanding the human experience in to space and further more a better understanding of our daily living conditions.

Acknowledgements

We would like to thank Tetsuya Kawabe and Madam Tomoko Kuwabara for knowledge about psychological issues. We also thank Tatsuhisa Hasegawa, Mahamadou Tandia, Prof. Kentaro Watanabe, and Prof. Satoshi Imura for helpful discussions.

Last but not the least, we are grateful to all our family members for sending us to Antarctica.

Author details

Giichiro Ohno[1,2*], Shinji Otani[3] and Atsushi Ikeda[4]

1 Department of Surgery, Tokatsu Hospital, Nagareyama, Japan

2 National Institute of Polar Research, Tachikawa, Japan

3 International Platform for Dryland Research and Education, Tottori University, Tottori, Japan

4 Department of Urology, University of Tsukuba, Tsukuba, Japan

*Address all correspondence to: oonog@mb.infoweb.ne.jp

References

[1] Hasegawa T, Oe H, Taki M, Sahkaguchi H, Hrano S, Wada Y. End-tidal CO_2 relates to seasickness susceptibility: A study in Antarctic voyages. Auris, Nasus, Larynx. 2017;**44**: 334-339

[2] Otani S, Ohno G, Shimoeda N, Mikami H. Morbidity and health survey of wintering members in Japanese Antarctic research expedition. International Journal of Circumpolar Health. 2004;**63**(Supp. 2):165-168

[3] Ohno G, Miyata T. Comparison of medical service systems at Syowa Station with other Antarctic stations: Medical staff, mortality and evacuation (in Japanese). Antarctic Record;**44**: 42-50

[4] Hasegawa Y, Watanabe K. International comparative study of medical service at Antarctic wintering-over stations (in Japanese). Antarctic Record. 2007;**51**:251-257

[5] Rogozov V, Bermel N. Auto-appendectomy in the Antarctic: Case report. BMJ. 2009;**339**

[6] Ohno G. Practical results of telemedicine system between Antarctic station and Japan. In: Graschew G, editor. Telemedicine Techniques and Applications. InTech; 2011. pp. 439–452. ISBN: 978-953-307-354-5. DOI: 10.5772/18474. Available from: http://dx.doi.org/10.5772/18474

[7] Ohno G, Miyata T. Morbidity of wintering-over participants in the first to thirty-ninth Japanese Antarctic Research Expeditions: Analysis of 4233 cases (in Japanese). Antarctic Record. 2000;**44**:1-13

[8] Otani S, Kusagaya M. Changes in cytokines at extreme surroundings in Antarctica. Yonago Acta Medica. 2003; **46**:29-34

[9] Otani S, Ohno G, Obinata I, Shimoeda N, Ohno H. Comparison of cardiorespiratory state between different approaches to a high altitude region in Antarctica (in Japanese). Mountain Medicine. 2006;**26**:87-90

[10] Ohno G, Miyata T. Diseases due to ultraviolet radiation in Antarctic wintering personnel: Analysis of genesis, seasonal change and annual variation (in Japanese). Antarctic Record. 2000;**44**:239-248

[11] Kawabe T, Naruiwa N, Shigeta T, Sasaki R, Kato N, Sasaki A, et al. Overview of and Outlook for Psychological Studies on Japanese Antarctic Research Expedition members. Human Sciences: Bulletin of Osaka Prefecture University. 2015;**10**: 123-141